A User's Guide

to the

Universe

Surviving the Perils of Black Holes, Time Paradoxes, and Quantum Uncertainty

Dave Goldberg and Jeff Blomquist

WILEY

John Wiley & Sons, Inc.

Copyright © 2010 by Dave Goldberg and Jeff Blomquist. All rights reserved

Published by John Wiley & Sons, Inc., Hoboken, New Jersey
Published simultaneously in Canada

Photo credits: page 48, © Akira Tonomura; page 187, Andrew Fruchter (STScI) et al., WFPC2, HST, NASA; page 204, NASA/WMAP Science Team; page 231, J. R. Gott & L.-X. Li

For general information about our other products and services, please contact our Customer Care Department within the United States at (800) 762-2974, outside the United States at (317) 572-3993 or fax (317) 572-4002.

Wiley also publishes its books in a variety of electronic formats. Some content that appears in print may not be available in electronic books. For more information about Wiley products, visit our web site at www.wiley.com.

Library of Congress Cataloging-in-Publication Data:
Goldberg, Dave, date.
A user's guide to the universe: surviving the perils of black holes, time paradoxes, and quantum uncertainty / Dave Goldberg and Jeff Blomquist.
    p. cm.
  Includes index.
  ISBN 978-0-470-49651-0 (cloth)
1. Physics—Popular works. I. Blomquist, Jeff II. Title.
QC24.5.G65 2010
530—dc22

2009028773

Printed in the United States of America

10 9 8 7 6 5 4 3 2 1

Contents

3 Randomness 67
"Does God play dice with the universe?"

4 The Standard Model 89
"Why didn't the Large Hadron Collider destroy Earth?"

5 Time Travel 131
"Can I build a time machine?"

6 The Expanding Universe 165
"If the universe is expanding, what's it expanding into?"

7 The Big Bang 199

"What happened before the Big Bang?"

8 Extraterrestrials 235

"Is there life on other planets?"

9 The Future 253

"What don't we know?"

Further Reading 281

Technical Reading 283

Index 291

Acknowledgments

This book has been a labor of love. We've tried to translate our love of teaching and our love of physics into something that could be understood and enjoyed by people at every level. We are so grateful for the feedback from our friends, family, and colleagues. First and foremost, Dave wants to thank his wife, Emily Joy, who was so supportive throughout, and who gave her honest opinions at every turn. Jeff wishes to thank his family (especially his brother), who remained politely neutral during the majority of his winded tirades and pointless doodling; he is also grateful to Frank McCulley, Harry Augensen, and Dave Goldberg, the three physicists who inspired him to give physics a fair shake. We are also indebted to feedback from Erica Caden, Amy Fenton, Floyd Glenn, Rich Gott, Dick Haracz, Doug Jones, Josh Kamensky, Janet Kim, Amy Lackpour, Patty Lazos, Sue Machler (aka Dave's mom), Jelena Maricic, Liz Patton, Gordon Richards, David Spergel, Dan Tahaney, Brian Theurer, Michel Vallieres, Enrico Vesperini, Alf Whitehead, Alyssa Wilson, and Steve Yenchik. We also would like to acknowledge Geoff Marcy and Evelyn Thomson, with whom we had several enlightening discussions. We appreciate Rich Gott and Akira Tonomura allowing us to reproduce their figures. Thanks also to our very hardworking agent, Andrew Stuart, and our excellent editors, Eric Nelson and Constance Santisteban.

Introduction

"So, what do you do?"

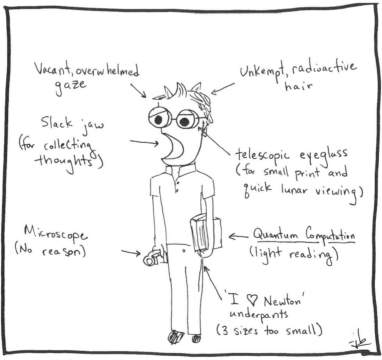

The 'Typical' Scientist

The life of a physicist can be a lonely one.

Imagine this: You sit down in an airplane, and the person next to you asks you what you do for a living. You reply that you're a physicist. From here, the conversation can go one of two ways. Nine times out of ten, the first thing out of his or her mouth is something along these lines: "Physics? I hated that class!"*

You'll then spend the rest of the trip (or party, or elevator ride, or date) apologizing for the emotional trauma that physics has apparently inflicted on your erstwhile friend. These random encounters often reveal an almost joyful contempt, reserved specifically for the fields of physical science and mathematics. "Oh, I'm terrible at algebra!" for example, is said in an almost boastful tone, in a way that "I barely even know how to read!" never would. But why?

Physics has a somewhat unfair reputation for being hard, impractical, and boring. Hard? Perhaps. Impractical? Definitely not. Indeed, when people try to "sell" physics to the public, it is almost always in terms

---

*On reading a draft of the manuscript, Mrs. Goldberg finally revealed to me that she was barely able to suppress a comment to this effect on our first date.

of how it can be used to build bridges or launch rockets—that is, how physics is ultimately the foundation for engineering or chemistry.

But boring? That's where we really take issue. The problem, as we see it, is that the practical side of physics is almost always put forward at the expense of the interesting side. Even folks with technical focuses such as engineering and computer science typically don't get past mechanics and electromagnetism to the really *fun* stuff. And that's a shame, because quite frankly there has been very little cutting-edge research done on pulleys in the past few years.

This hostility to physics seems to be ingrained, and makes it difficult to have discussions without jading an audience. In starting a scientific conversation with a "civilian," we purveyors of physics often feel like we're trying to force people to eat their vegetables, and rationalize it in the same way. We never begin physics discussions with "It's fun!" but almost always with "It's necessary," which naturally drains all of the fun out of it.

In an era when new technologies are constantly emerging, scientific literacy should be fundamental. On the other hand, it isn't necessary that you have four extra years of college sciences to understand them. You don't need to have a detailed knowledge of exactly how the physics works to appreciate the revolutions in quantum computing or cosmology. It is important, rather, to understand *why* these developments are significant, and how they are poised to change technology and our lives.

And it's not simply that people need to understand a particular theory. Physics is the archetypal inductive science, and by understanding how science proceeds, people are better able to make informed decisions about issues from global warming to "theories" of intelligent design. The hope is that we are more prepared to refute people who disagree with us by offering facts rather than simply insisting "No."

The United States, in particular, has an immense problem with science and mathematics education, with high school students performing well below average compared to those in other developed countries. But we cannot limit ourselves to *only* blaming teenagers, or their teachers, or, for that matter, programs such as No Child Left Behind.

The problem is far-reaching, affecting all walks of life. It is most evidently *manifested* in teenagers because we don't sit down with people in their fifties and ask science-y questions such as, "If you have ten chickens and you eat five of them, how much does your cholesterol go up?" Looking at a so-called practical story problem now makes the whole premise of applied math seem absurd. At a very early stage, many children throw up their arms and say, "When am I ever going to need algebra?" and assume that the sole virtue in studying for the class is getting a good grade.

In an excellent series of books, John Allen Paulos addresses the epidemic of "innumeracy" and through a series of lively essays on topics that students normally don't see, tries to give his readers the ability to think critically about numerical concepts, and tries to show (successfully, in our opinion) that mathematics is interesting above and beyond its practical import in computing the tip on your bill or balancing your checkbook.

As your own experience may suggest, physics has the same break between the practical and the groundbreaking. Although dry, mechanics-based classes may drive people away from physics, they are sometimes drawn back in by science fiction, or newspaper accounts of big discoveries, or the latest pictures from the Hubble Space Telescope.

These accounts, however, rarely feature the latest breakthroughs in inclined plane technology.

Rather, when the public gets excited, it tends to be about the universe, or big experiments such as the Large Hadron Collider, or life on other planets. We said before that nine times out of ten, our attempts at discussing physics at an airport or cocktail party left us with no phone number and a lonely cab ride home, but the rest of the time something wonderful happens. Occasionally we will actually have *conversations* instead of *confrontations*. Sometimes we're lucky enough to be seated next to somebody who had a great physics teacher in high school, or whose uncle works for NASA, or who is an engineer and thinks what we're doing is simply "quaint."

In these cases, the conversation goes quite differently. It seems that every so often we run into someone who has been holding a question about how the universe works in reserve for some time but couldn't figure out the keywords to plug into Wikipedia. Maybe the latest *NOVA* special only hinted at a topic, and they were eager to know more. Some recent questions have included:

- I heard that the Large Hadron Collider is going to create mini black holes that will destroy the universe. Is this true? (Providing yet more evidence, as if any were needed, that physicists are perceived as nothing more than mad scientists who would love nothing more than to destroy Earth.)
- Is time travel possible?
- Are there other, parallel universes?
- If the universe is expanding, what's it expanding into?
- What happens if I'm traveling at the speed of light and I try to look at myself in the mirror?

These are the sorts of questions that got us excited about physics in the first place. Indeed, the last question on the list above was one that Albert Einstein himself posed, and was one of the main motivations for his development of special relativity. In other words, when we talk to people about what we do, we find that some people, however rare they may be, are excited about exactly the same aspects of physics as we are.

The most obvious method is to make the subjects more approachable through available mathematics and science teaching materials. In response to this, most textbook authors try to make physics exciting by putting pictures of volcanoes, locomotives, and lightning bolts on the covers.* The desired response, presumably, is that students will look at the book and say, "Cool! Physics is really coming alive for me!" Our own experience is that students aren't fooled by these ploys. If they are, they end up looking for the "How to Make Your Own Lightning" chapter, and are even more disappointed when they fail to locate it.

We'd like to note in passing that we don't take that approach in this book. You won't see any cool graphics,† or anything else likely to increase the publication costs of the book. Rather, our approach will be quite simple: the physics itself is interesting. No, really! And if you need further persuasion, we solemnly promise to deliver no fewer than five bad

---

*On one humorously misguided cover, a bowling ball striking pins was intended to "bowl students over" with the power of physics.

†Though at least one of the authors submits that all the drawings are witty, informative, or both.

jokes per chapter (including groaners, puns, and facile cartoons). To give you an idea of the sort of family-friendly humor you're in for, consider the following:

Q: What did the photon do at the ballpark?

A: The lightwave!

With that in mind, each chapter of this book will start with a cartoon featuring an inexcusably terrible pun, and a question about how the universe works. By way of answering the question, we're going to take you on a tour of the physics surrounding it, and by the end of the chapter, it's our hope that the mystery surrounding the question will become clear, and that given the opportunity to reexamine it, you will find the cartoon hilarious. We will do so in exactly the way you'd expect from scientists—very circuitously.

That is not to say that you must be a physics guru to understand; quite the contrary. Our aim is to find some middle ground between those who appreciate the underlying majesty of the physics foundation and those who would rather gag themselves with a spoon than be caught dead within a hundred yards of a protractor.

Without equations, many science writers usually resort to analogies, but the problem is that it isn't always clear to the reader that what's being written is an analogy rather than a literal description of a problem. Without using math, it's clear that there will be some crucial element of the physics missing. What we'd like to convey is how you would want to *think* about the problem, even if you don't have the equations to set it up. In other words, once you understand what's really going on, doing the math is just, well, math.

This description raises this question: *What exactly do you eggheads expect from me?* In writing this book, we make no presumptions. Every bit of evidence we present is constructed from the basics. It is not our intention to scare you with mathematics or daunting equations. In fact, why don't we get all of the equations out of the way right now?

$$E = mc^2$$

That's it. That didn't hurt too badly, did it?

1

Special Relativity

"What happens if I'm traveling at the speed of light, and I try to look at myself in a mirror?"

A photon is grilled to recall the events of the last hundred years.

All high school experiences have one thing in common: there are always a handful of students—*the cool kids*—who feel the insatiable need to mock everything and everyone around them. This is why we like to think of ourselves as the *cool kids of physics*, if such a thing could be said to exist. We'll give you an example.* We spent part of the introduction making fun of textbook authors who need to use examples involving cataclysmic natural events, sports, or monster trucks to "make physics come alive." We aren't backpedaling, but some of those goofy examples have a tiny bit of merit.

That, and we know in our heart of hearts that we'll never get this physics party started unless we set off some fireworks. If you've ever been to the local Chamber of Commerce Independence Day celebration and decided to get a little physics in, you'll have noted that there's a time delay between the rockets' red glare and the sounds of bombs bursting in air. You see the explosion several seconds before you hear the sound. You've probably experienced the same thing if you've ever had back-of-the-theater tickets at a concert: the music and the musicians suffer a delay. Sound moves fast, but light moves faster.

---

*And, perhaps, a wedgie.

In 1638, Galileo of Pisa (one of the *original* cool kids of physics) devised a scheme to figure out the speed of light. The experiment went like this: Galileo parked himself on a hill with a lantern, while his assistant, armed with his own lantern, walked far away to a different, distant hill. The two signaled each other. Each time Galileo saw his assistant's lantern open or close, he would toggle his own, and vice versa. By performing the experiment on more and more distant hills, Galileo hoped to measure the speed of light. The precision wasn't really there, but no one can blame him for taking a crack at it, and he did come to a pretty interesting conclusion.

If it isn't infinite, the speed of light is pretty darn fast.

Galileo concocts the perfect scheme
to ditch boring dates.

Over the next few centuries, physicists made ever more precise measurements, but we won't bother you with the design specs for the intricate instrumentation. Suffice it to say that as time went on, scientists grew more and more determined to shed light on light.

The modern value of the speed of light is 299,792,458 meters per second. Rather than rattle off all of the digits, we'll simply call it *c* for the Latin *celeritas*, meaning "swift." This measurement is not the kind of number you get with a ruler and an egg timer. To measure *c* this precisely, you have to use an atomic clock powered by cesium-133 atoms. The scientific community defines the second as *exactly* 9,192,631,770 times the frequency of light emitted by the "hyperfine transition" of cesium-133. This may sound like it's unnecessarily confusing, but it actually simplifies things a great deal.* The second, like your hat size, becomes something that we define in terms of something real; a bunch of physicists could build cesium clocks, and since all cesium acts the same, everyone tells the same time.

We've come up with a creative way of defining the second, but how does that help us measure the speed of light? Speeds are ratios of distance over time, such as *miles per hour*, and defining the second gives us some leverage. The only thing left to do is determine the length of a meter. This may seem pretty obvious since a meter is exactly one meter long. Just get out a meter stick and you're all set. But how long is that?

From 1889 until 1983, if you wanted to know how tall you were, you'd have to go to the International Bureau of Weights and Measures in Sèvres, France, go into their vault, and take out their platinum meter stick to measure yourself. Not only was this cumbersome (and illegal, if you didn't ask nicely to use it first), it tends to be pretty inaccurate. Most materials, including platinum, expand when heated. Under the old system, a meter was slightly longer on hot days than cool ones.

So instead of using an actual meter stick, we have a clock capable of measuring a second, and we *define* a meter as 1/299,792,458 the distance that light travels in 1 second. To make this blindingly obvious, what we've done is say, "We know the speed of light *exactly*. But meters, on the other hand, have a tiny uncertainty." All this hard work means that we can normalize the second and the meter, and everyone uses the same measurement system.

---

*At least it simplifies things for scientists who know what "hyperfine transitions" are. You don't need to know; it won't be on the test.

Keep in mind, though, that the crux of it all is that light doesn't move infinitely fast. Not impressed? Brace yourself for a philosophical bombshell: because light moves at a finite speed, we are forever gazing into the past. As you're reading this book, a foot in front of you, you're seeing it as it was about a billionth of a second earlier. The light from the Sun takes about eight minutes to reach Earth, so our star could well have burned out five minutes ago and we'd have no way of knowing it.[*] When we look at stars in our Galaxy, the light takes hundreds, or even thousands of years to reach us, and so it is a very real possibility that some of the stars we see in the sky are no longer around.

Why can't you tell how fast a ship is moving through fog?

No experiment has ever produced a particle traveling faster than the speed of light.[†] The speed limit of the universe seems to be something we can't brush off even if we wanted to, and the constant speed of light is just the first of two ingredients in what will turn out to be one of the finest physics dishes ever cooked. For the second, we need to think about what it even means to be moving at all.

Allow us to introduce you to Rusty, a physicist-hobo riding the rails, ostracized by society for the unique standards of hygiene common to his lot. Rusty has managed to "borrow" the platinum meter stick from the International Bureau of Standards (which, while not perfect, is still *pretty good* by hobo standards), and he has a bunch of cesium atoms to build an atomic clock.

He passes his day by throwing his bindle[‡] across the train. Each time he throws it, he measures the distance it travels, and the time it takes

---

[*]At least not for another 180 seconds or so.

[†]For those of you especially well versed in sci-fi lore, you might have heard of a hypothetical particle called the tachyon, which can *only* travel faster than the speed of light. No one has ever detected one. As a real particle (rather than as a mathematical construct), the tachyon is really most at home in science *fiction* rather than in this discussion.

[‡]In case you've forgotten, that's the stick with a polka-dot sack at the end of it.

to cover that distance. Since speed is the ratio of distance traveled compared to the time it takes to cover that distance (miles per hour), Rusty is able to calculate the speed of his bindle with high accuracy.

After a tiring day of bindle-tossing, Rusty nods off to sleep, and he awakes in his own private freight car. Since freight cars don't have any windows, and the train is moving on smooth track, he finds himself somewhat disoriented when he slides open the door and finds that he is moving. You may have noticed that even in cars, you sometimes can't tell that you're moving without looking out the window.

You also may not have noticed that if you're standing on the equator, you're moving at more than 1,000 mph around the center of Earth. Faster still, Earth is moving at about 68,000 mph around the Sun. And the Sun is moving at close to 500,000 mph around the center of our Milky Way Galaxy, which, in turn, is traveling through space at well over 1 million mph.

The point is that you (or Rusty) don't notice the train (or Earth, or the Sun, or the Galaxy) moving, regardless of how fast it's moving, as long as it does so smoothly and in a straight line.

Galileo used this argument in favor of Earth going around the Sun. Most people at the time assumed that you'd be able to somehow *feel* Earth's motion as it flies around the Sun, so therefore we must be standing still.

"Nonsense!" said Galileo. Not having a ready supply of either hobos or trains, he compared the motion of Earth to a ship moving on a calm sea. It's impossible for a sailor to tell under those circumstances whether he's moving or standing still. This principle has come to be known as "Galilean relativity" (not to be confused with Albert Einstein's special relativity, which we will encounter shortly).

According to Galileo (and Isaac Newton, and ultimately Einstein) there is quite literally no experiment you can do on a smoothly moving train that will give a different result than if you were sitting still. Think back to trips with your family in which you threw mustard packets at your little brother until your parents threatened to "turn this car around this minute, young man!" Even though the car was moving at 60 mph or more, you threw the packets exactly as you would have if the car were sitting still. Like it or not, all of that tormenting was nothing more than a simple physics experiment. On the other hand, this is only true if the speed and direction of the car/train/planet/galaxy are exactly (or really, really close to) constant. You definitely felt it if your parents actually made good on their threat and slammed on the brakes.

So when he awakes from his blissful hobo slumber to return to his bindle-tossing experiments, Rusty might be quite unaware that the train has started steadily moving at about 15 mph. After arranging himself at one end of the train car, he tosses his bindle and measures the speed at, say, 5 mph. Patches, a fellow hobo-physicist, stands outside the moving train but also decides to participate. Using special hobo X-ray goggles to see through the train's walls, he also measures the speed of the bindle as Rusty throws it. Patches, from his vantage point outside the train, finds the bindle to move at about 20 mph (the 15 mph that Rusty's train is moving plus the 5 mph of the bindle).

So who's right? Is the bindle moving at 5 mph or 20 mph? Well, both are correct. We'd say that it's moving at 20 mph *with respect to Patches* and 5 mph *with respect to Rusty*.

Now imagine that our train has a high-tech lab equipped with lasers (which, being made of light, naturally travel at *c*). At one end of the train sits the laser, manned by Rusty. At the other end of the train sits an open can of baked beans. If Rusty turned on the laser for a short pulse (to heat his baked beans, naturally) and measured the time for the beans to start cooking, he could compute the speed of the laser, and he'd find it to be *c*.

B(mc²)ANS

What about Patches? He will, presumably, measure the same amount of time for the light pulse to reach the detector. However, according to him, the light doesn't have to travel as far to get there, so he should measure the speed of the pulse to be faster than *c*. In fact, common sense tells us that he should measure the pulse to be moving at *c* + 15 mph.

Earlier we said that Einstein assumed that the speed of light is constant for all observers, but by our reasoning the beam doesn't appear to be constant. Not constant at all! Could the great Einstein be wrong?[*]

Fifteen pages into the book, and we've already broken the laws of physics. We couldn't be any more embarrassed if we showed up to a party wearing the same dress as the hostess. It looks like we just blew it. If only there were some obsessive scientist we could look to, some concrete example to revalidate the concept of c as a constant.

We just so happen to have such a scientist. His name was Albert Michelson, and he loved light in a way that today might be characterized as "driving" or "unhealthy." His scientific career began in 1881, after he left the navy to pursue science. He measured light independently for a while, doing gigs in Berlin, Potsdam, and Canada, until he met Edward Morley. They worked together to produce ever more elaborate devices for measuring the speed of light, eventually reaching number 1 with "Bridge over Troubled Water," which stayed at the top of the charts for six straight weeks.

The devices they constructed worked on the following basic premise: since Earth travels around the Sun once a year, relative to the sun their lab should travel at different speeds and in different directions at different times of year. Michelson's "interferometer" was designed to measure whether the speed of light was different when moving in different directions. Your basic intuition should tell you that as Earth moves toward or away from the Sun, the measured value of c should change.

Your intuition is wrong. In experiment after experiment, Michelson and Morley showed that no matter what the direction of motion, the speed of light was the same everywhere.

As of 1887, this was a pretty big conundrum, and it defied the senses because this only seems to work for light. If you found yourself on a bike, face-to-face with an angry cow, it would make all the difference in the world whether you rode toward or away from the charging animal. Whether you run toward or away from a light source, on the other hand, c is c.

---

[*]Not in this case. But he did mess up at least twice, and we'll talk about those instances in chapters 3 and 6.

Putting it even more bluntly (on the off chance that the strangeness of this still isn't clear), if you were to shine a laser pointer at a high-tech measuring device, then you would measure the photons (light particles) coming out of the laser pointer at about 300 million meters per second. If you were in a glass spaceship traveling away from a laser at half the speed of light (150 million meters per second) and someone fired the laser beam through your ship to a detector, you would *still* measure the beam to be traveling at the speed of light.

How is that even remotely possible?

To explain this, we need to take a closer look at a hero of physics, the "Light"-Weight Champion[*] of the World: Albert Einstein.

How fast does a light beam go if you're running beside it?

When Einstein first proposed his principle of special relativity in 1905, he made two very simple assumptions:

1. Just like Galileo, he assumed that if you were traveling at constant speed and direction, you could do any experiment you like and the results would be indistinguishable from doing the same experiment in a stationary position.

    (Well, sort of. Our lawyers advise us to point out that gravity accelerates things, and special relativity relies on there being no accelerations at all. There are corrections that will take gravity into account, but we can safely ignore them in this case. The correction required for the force of gravity on Earth is very, very small compared to the correction near the edge of a black hole.)

2. Unlike Newton, Einstein assumed that all observers measure the same speed of light through empty space, regardless of whether they are moving.

---

[*]Get it?

In our hobo example, Rusty threw his bindle and measured the speed by dividing the length of the car by the time the bindle took to hit the side. Patches sat by the side of the tracks and watched the train and bindle speed by, and therefore saw the bindle move farther (across the car and across the ground the car covered) in the same amount of time. He saw the bindle move faster than Rusty did.

But now consider the same case with a laser pointer. If Einstein was right (and Michelson and Morley's experiment demonstrated, almost two decades earlier, that he was) then Rusty should measure the laser moving at c and Patches should measure the *same exact speed*.

Most physicists believe that c is a constant without batting an eyelash, and use it to their collective advantage. As a form of exploitation, they frequently express distances in terms of the distance light can travel in a particular amount of time. For example, "light-seconds" are approximately 186,000 miles, or about half the distance to the Moon. Naturally, it takes light 1 second to travel 1 light-second. Astronomers more commonly use the unit "light-year," which is about 6 trillion miles—about a quarter the distance to the nearest star outside our solar system.

So let's make our previous example a little weirder and give our hobo-physicist an intergalactic freight car. It's 1 light-second long, and while Rusty has more space than he will ever need to stretch out and nap, he has the perfect amount of space to run his laser experiment again. He fires off the laser from the back of the train and, by his reckoning, the laser takes 1 second to traverse the train. It must, after all, because light travels at the speed of light (duh!).

But Patches watches the light beam on the moving train and says (correctly) that while the beam was traveling, the front of the train moved farther ahead, and therefore, according to Patches, the beam traveled farther than measured by Rusty's reckoning. In fact, he finds that the beam travels a total of 1.5 light-seconds. Since light must still travel at the speed of light, Patches will find that it takes the light pulse 1.5 seconds to go from the laser to the target.

Let's be clear: Rusty says a particular series of events (the pulse being shot and then hitting the target) takes 1 second, and Patches says that the same series of events takes longer. Both have perfect working

watches that were built at the same intergalactic hobo-physicist depot. Both made excellent measurements. Who's right?

They both are.*

No, really. If the speed of light is the same for both Rusty and Patches, then Patches *must* interpret what he sees by saying that his own clock must be fast—or Rusty's clock was running slow. The weirdest part is that this is true of every clock in Rusty's train. He sees pendulums swinging slowly, wall clocks ticking slowly, and even (if he had the equipment to measure it) old Rusty's heart pumping away more slowly than usual.

---

*Whaa . . . ?

This is true in general. Whenever you see someone speed by you, their clocks will run more slowly as far as you're concerned, but you don't have a watch precise enough to show this. If you look overhead and see a plane flying by at about 600 mph, and somehow you had the keen eyesight to see the captain's watch, you could see her clock running slower than yours—but only by 1 part in about 10 trillion! In other words, if the captain flew for 100 years, by the end of that period, she would have escaped from almost an entire second's worth of aging. So even though this effect (called "time dilation") is always in force, the fact is that you will never notice it in your everyday life.

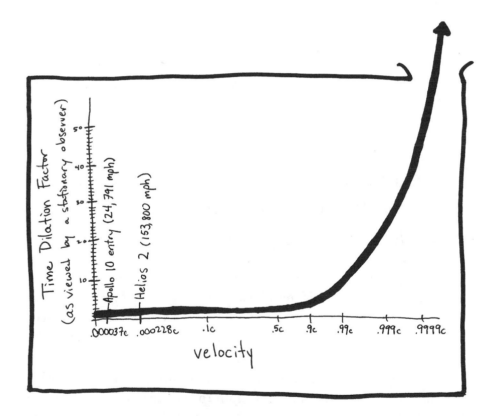

Time dilation really kicks in when you start going close to the speed of light. We're not going to give you the exact equation, so you'll have to take our word for it that we're doing the calculations correctly. If the train were going half the speed of light, then for every second on Rusty's clock, 1.15 seconds would pass on Patches'. At 90% the speed of light, for every one of Rusty's seconds, Patches would measure 2.3 seconds. At 99% the speed of light, the ratio becomes 7:1. And as the speed gets closer and closer to c, the number gets bigger.* The time dilation factor becomes infinite as the train gets to c—which is our first hint that you can't actually move at the speed of light.

It's not just time, either. Space behaves the same way. Let's imagine that Rusty is ramblin' on down the track toward a switching station at a sizable fraction of the speed of light. Let us also imagine that Patches is trying to sleep at the same switching station. Rusty covers the distance along the ground in a shorter amount of time by his own reckoning than by Patches'. Since they both agree that the train is approaching the station at the same speed, Rusty must think that the total distance to the station is shorter.

Time and space really are relative to your state of motion. This is not an optical illusion; it is not a psychological impression; it is actually how the universe works.

If you head off in a spaceship traveling at nearly the speed of light, what horrors await you when you return?

While this might seem like trifling over vague curiosity, scientists have figured out ways to exploit this phenomenon for more interesting study. As an example of the sort of grand pronouncements we can now make about the universe, consider the humble muon. Never heard of it? We don't blame you. If you have a muon, then you'd better treasure your time together, because, on average, they last only about a millionth of a second (the time it takes a light beam to travel about half a mile, or the total duration of Vanilla Ice's acting career) before they decay into something else entirely.

---

*As does the likelihood that Rusty will step off his boxcar into a world populated by super-intelligent, damn, dirty apes.

Between how they're made and how long they stick around, there aren't a heck of a lot of muons around. They primarily form when cosmic rays hit the upper atmosphere and create particles called pions (which are even shorter-lived) and then those pions decay into muons. This all happens about 10 miles out from the surface of Earth. Since nothing can travel faster than light, you might suppose that the farthest muons can travel before decaying is about half a mile and that none of the muons will reach the ground.

Once again, your intuition is not quite right.* The muons have such high energy that many of them are moving 99.999% of the speed of light, which means that to us on the ground, the "clocks" inside the muons—the very things that tell them when to decay—are running slow by a factor of about 200 or so. Instead of going half a mile without decaying, they are able to go 100 miles before decaying, easily enough to reach the ground and then some.

Perhaps a scenario that will make a bit more sense involves the so-called twin paradox. There are twin sisters, Emily and Bonnie, who are thirty years old. Emily decides to set out for a distant star system, so she gets in her spaceship and flies out at 99% the speed of light. After a year, she gets a bit bored and lonely and returns to Earth, again at 99% of the speed of light.

But from Bonnie's perspective, Emily's clock—and watch, and heartbeat, and everything else—have been running slow. Emily hasn't been gone for two years; she's been gone for fourteen! This is true however you look at it. Bonnie will be forty-four; Emily will be thirty-two. You can even think of traveling close to the speed of light as a sort of time machine—except it only works going forward and not backward.

There are other, perhaps subtler effects as well. For example, since Emily was traveling away from Earth for seven years (according to Bonnie) at nearly the speed of light, she must have gotten 7 light-years from Earth before turning tail and returning. This takes her most of the way to Wolf 359, the fifth-nearest star to our Sun. By Emily's account, though, she knows that she can't travel faster than light, so in her 1 year

---

*You're still going home with this book as a consolation prize. And unless someone is reading over your shoulder, only you know what a terrible guesser you are.

outbound, she'll say that her distance traveled was only 99% of a light-year. In other words, while on her journey, she measures the distance between the Sun and Wolf 359 to be only about 1 light-year.

This effect is known as "length contraction." Like with time dilation, length contraction isn't just an optical illusion. While she is traveling at 99% the speed of light, Emily measures everything to be shrunk along her direction of motion by a factor of 7. Earth would appear squashed, and Bonnie would appear to be rail thin as well, but with her normal height and breadth.

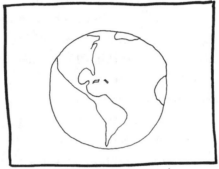

EARTH : NORMAL VIEW

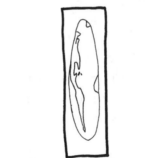

EARTH: IF IT IS NOT A CIRCLE,
YOU ARE READING TOO QUICKLY

Like with time dilation, we don't notice this effect in everyday life. If our pilot friend took the time to look down from her plane, the streets below would seem slightly thinner than normal, but even flying at 600 mph, the difference amounts to about 0.04% the size of an atom. While relativity is useful for explaining bizarre and interesting high-speed phenomena, it is clear that it is a poor excuse for a healthy diet and exercise.

The time dilation and length contraction *should* be observed symmetrically when Bonnie is looking at Emily or Emily is looking at Bonnie. Here's where the paradox comes in. When Emily steps off her ship back on Earth after traveling to Wolf 359 and back, everyone agrees that she's aged only two years in the same time that Bonnie has aged fourteen. That is totally *inconsistent* with pretty much everything we just told you, because we immediately know that Emily was the one who "moved" and not Bonnie, and the first rule was that you could never tell who was moving and who was sitting still. So how do we resolve it?

There is one rule we gave you early on that tells you whether special relativity is the law of the land—for special relativity to work, you need to be moving at constant speed and direction. And to move things along, we'll tell you that Emily certainly wasn't. She had to launch her ship to get off Earth and get up to speed (during which she felt a tremendous force of acceleration), she needed to decelerate and reverse direction when she reached Wolf 359, and then she needed to slow down to land when she got back to Earth.

With all of those accelerations, all bets are off, and we need a much more complicated theory to describe everything. To put things in a bit of historical perspective, Einstein came up with his theory of special relativity (no accelerations) in 1905, and didn't get the theory of general relativity right (which includes gravity and other forms of acceleration) until 1916.

Can you reach the speed of light (and look at yourself in a mirror)?

We've taken a heck of a digression from our original question, and that's a shame, because it's a good question—so good, in fact, that it's the very one Einstein asked himself. You may feel, however, that we're no closer to answering the question than we were before.

Au contraire![*]

Our answer will actually have two parts, and one of them you're already prepared to answer (and have been for some time). Think back about

---

[*]Tr.: "Don't touch that dial!"

old Rusty in his train. Now imagine that Rusty's train is traveling at 90% of the speed of light (or any other speed you like). Rusty, however, is unaware of anything around him because he's too busy preening for his date with Hambone Lil. As Rusty gazes into the reflection of his handsome mug, does he see anything amiss? He does not. Since there are no windows in his boxcar, and he's moving on straight, smooth track, there is no experiment he can do that shows he is moving rather than sitting still. As long as the mirror is moving with Rusty, he looks the exact same as he would were he not on the train.

All of this is fine and good if Rusty is traveling slower than light, but what if he's traveling at the speed of light? We know, we know, we've said that nothing can travel at the speed of light, so perhaps you'll be inclined to just take that at face value. But why should you?

We can illustrate. Patches, jealous of Rusty's success with the ladies, watches Rusty prepare for his date. Of course, he has to pay very keen attention, as Rusty's train is speeding by at 90% of the speed of light. Tragedy strikes for Rusty, who gets a call from Lil, who is phoning to cancel. She lets him down easy, but Rusty is still upset, and thus picks up his still-warm can of beans and hurls it toward the front wall at 90% the speed of light (as seen by him).

Patches may be overcome with schadenfreude, but he's not too distracted to note how fast the can of beans is flying from his own perspective. Now, in his own naive youth, he might have assumed that the beans were moving at $1.8c$—the speed of the train ($0.9c$) plus the speed of the beans within the train ($0.9c$). But he has long since left behind that sort of foolishness.

Remember the two facts:

1. He sees Rusty's clock running slow (in this case, by a factor of 2.3).
2. He sees Rusty's train compressed (again, in this case, by a factor of 2.3).

The details obviously don't matter too much, but the important thing is that according to Patches:

1. The beans take a far longer time to go from Rusty's hand to splattering against the wall than Rusty says they do.
2. The beans don't travel nearly as far as Rusty says they do.

The point is that the beans are going far slower than our (and Patches') original naive estimate. Instead of 1.8c, the beans are moving a paltry 99.44% the speed of light.

We could keep playing this game indefinitely. For example, imagine that there was an ant sitting on the can. The ant had big plans with the queen of his colony until she called to inform him that she had to stay in to clean her thorax. In anger, he threw a crumb of food at 0.9c (from his perspective) toward the front of the train. Patches, with his unbelievably keen eyesight, would see the crumb moving at 99.97% of the speed of light.

And if on the crumb there lived an amoeba who, reproducing asexually, stood itself up for a date . . . you get the picture.

No matter how hard we try, no matter how many boosts we give to something, we can't ever get it going up to the speed of light. It just gets closer and closer and closer.

It also requires more and more work to get things moving faster as it gets closer and closer to the speed of light. It seems that it would take twice the work to get something moving at 99% of the speed of light compared to 50% of the speed of light; in fact, it takes more than six times as much work. And it takes more than three times as much work to get up to 99.9% of the speed of light from only 99%.

So now we can work up to the question posed by sixteen-year-old Einstein[*]: What happens if you travel at 99% of the speed of light and look at yourself in a mirror? Nothing, or at least nothing unusual. Your spaceship looks normal; your internal clocks seem to run normally. Your mug looks exactly as it always has. The only thing that you might notice is that your friends back at home see their hearts, clocks, cheese-cake calendars, and every other assorted timepiece running about seven times slower than they should. Also, for some reason, they appear to be smooshed by the same factor.

We could take it a step further and ask if anything appears amiss to someone looking in the mirror and traveling at 99.9% of the speed of light. The time dilation and length-contraction numbers are a bit bigger (a factor of 22 rather than 7), but otherwise everything's the same.

---

[*]Or at least the question we know about. Kids can be very curious at that age.

The problem here is that each of these speeds, while very, very close, is still less than the speed of light. Every tiny incremental speedup requires more and more energy, but to actually get up to c would require an infinite amount of energy. Not very big, mind you. Infinite.

Perhaps you're not satisfied with that. *If* you could somehow go at the speed of light (never mind that it's impossible), the light from your face could never reach the mirror, and therefore, much like a vampire, you wouldn't be able to see your reflection. But wait! The very fact that you wouldn't see a reflection would make it immediately obvious to you that you were going at the speed of light. But since we've already determined that nobody can ever tell that they are the ones in motion, this proves that you cannot get up to the speed of light.

Isn't relativity supposed to be about turning atoms into limitless power?

All of this about clocks and meter sticks and the speed of light may be interesting enough in their own right, but they're probably not the first things you think of when (and if) you think about relativity. You almost certainly think about the most famous equation in all of physics (and the only one we're going to write out explicitly in this book):

$$E = mc^2$$

Writing it out is simple enough, and by now you're even familiar with one of the terms in the equation: c, the speed of light.

The E on the left stands for energy, and in a moment we'll talk about how energy enters into it, but for now we're going to focus on the other term, m, which stands for mass.

You may think of mass as a measure of the "bigness" of a thing, but to a physicist mass is simply how hard it is to get something moving and how hard it is to stop it once it's moving. It's far easier to stop Rusty when he's running at you at 10 mph than it would be to stop his train moving at the same speed.

But we've already noticed something interesting about the effective mass of, in our case, a can of beans. We found that as the speed of the can of beans gets higher and higher, it requires more and more work to speed it up even a little bit. In other words, the beans and the can act as if they are getting more and more massive (that is, harder and harder to move). And, as we already observed, if the speed of the can gets arbitrarily close to the speed of light, eventually you need to do an infinite amount of work to speed the can up at all.

Put another way, as the energy of motion increases, the *inertial mass* seems to increase as well; that is to say, the can does not acquire more matter, but it behaves as if it does. But even if the speed of the can goes down to zero—which is to say that there is no energy of motion—the inertia of the can doesn't go away. If the can and the beans are completely stationary, they have a certain amount of energy, a sort of *minimum* inertial mass. The inertial mass can only increase from here as energy is added.

Einstein's famous equation is really a conversion formula between mass and energy.

The formula has a plethora of interesting applications, and we quite literally see the repercussions of it every second of every day of our lives in the radiation from the Sun. Even with the seemingly successful application of Einstein's theory, though, there has been an incredible impact on popular perception, especially by those who do not understand it.

As a working scientist, one of your esteemed narrators (Goldberg) frequently gets manuscripts from people with claims that they have a theory that will overturn the existing paradigm of science as we know it, and nine times out of ten, the central thesis of their argument is that Einstein's great equation was wrong, that there was some flaw in his reasoning, or that the math simply admits of an alternative explanation. This phenomenon is so pervasive (and ongoing) that a hundred years after Einstein first derived his equation, the NPR program *This American Life* did a story on a man who tried (unsuccessfully) to show that "*E* does not equal *mc* squared."

Why does this fascination with a simple conversion exist? In part, it's because the equation looks so simple. There are no unfamiliar symbols, and most people have a working understanding of all of the terms in the equation. And in a real respect, the equation *is* simple. It's a way of saying, "I'd like to trade in my *stuff* for energy. What'll you give me for it?"

The answer is "rather a lot." The reason is that we've already established that *c* is a big number, and we multiply the mass by the square of *c* in order to calculate the energy released.

We'll start small. Let's say that you have about 2 grams of boomonium, a substance we just invented just so we could use the name. The amount you have is about the mass of a penny, and you somehow manage to convert it all to energy. Were this possible—and we assure you it is not—you'd get out about 180 trillion joules of energy. Don't have an intuitive feel for how much that is? No problem. With the energy released you could:

1. power more than fifty-thousand 100-watt lightbulbs for a year;
2. exceed the caloric energy consumption by the entire population of Terre Haute, Indiana (pop.: 57,259), for a year; or
3. equal the energy output of about five thousand tons of coal or about 1.4 million gallons of gasoline. Provided they carpooled, this would be enough to drive everyone in Terre Haute from New York to California. It is not clear, however, why you would want to do that.

By comparison, the normal combustion energy of 2 grams of coal can power one lightbulb for about an hour.

Like most people, matter doesn't always live up to its full potential, and with the exception of cases where we smash matter into antimatter (which we shall return to), there is nothing that converts all of its mass into energy. So before you assume that it's just a quick step from $E = mc^2$ to complete energy independence from oil, hold on.

Einstein's famous equation changed the world, with the most obvious examples being the development of nuclear weapons and nuclear power. It's important to recognize that in most nuclear reactions, we convert only a small fraction of the total mass of a material into energy. The Sun

"According to my calculations, you have PLENTY of Energy!"

is a giant thermonuclear generator that turns hydrogen into helium. The basic reaction involves taking 4 hydrogen atoms and turning them into 1 helium atom—plus some waste products, including neutrinos; positrons; and, of course, energy in the form of light and heat. This is great news for us, since the energy produced by the Sun is collected as light rays, warms the surface of Earth, feeds algae and plants, and ultimately sustains us as an ecosystem.

However, it's not nearly as efficient as our boomonium. For every kilogram of hydrogen that is "burned" by the Sun,* we get 993 grams of helium back, which means only 7 grams get converted into energy. Still, as we've already seen, a little mass goes a long way.

The most common examples of mass-energy conversion come in the form of turning mass into energy rather than the other way around,

---

*Physicists like to point out that nuclear reactions aren't really burning. Burning is a chemical process, not a nuclear one, and requires oxygen to run. We are a very pedantic bunch.

including some of the scarier stuff out there: nuclear bombs, power plants, and radioactive decay. In each of these cases, a high-energy collision or random decay forces a small amount of mass to be converted into a walloping huge amount of energy. Why are radioactive materials so scary? Because the energy produced by even a single decay produces a photon of enormous energy, enough to do serious damage to your cells if given half a chance.

In the very early universe, it was more often the case that energy became matter, though it rarely happens anymore. At that time, when temperatures were billions of degrees, matter actually came out of light particles smashing into each other. Sound fascinating? It sure does. And that's why we'll return to it in chapter 7.

Physics Smackdown: Who Is the Greatest Physicist of the Modern Era?

TOP FIVE

Every now and again, we get drawn into inane discussions at the level of "Who's better: Kirk or Picard?" or "Who is the best physicist?" While the former should be obvious to anyone who isn't a *yIntagh*,* the latter is just way too vague. For our money, we'd argue that the greatest physicists are those who have something really important named after them—even if someone else came up with it independently. Sometimes, great thinkers don't get the credit they deserve (we're thinking of you, Tesla), but for the purpose of our list, that's just their bad luck. Also, because we want to keep things fresh, we're afraid that everybody who did their best work before 1900 is shut out. Finally, we're sure that there are lots of physicists who would disagree with our list, and to them, we respectfully suggest that they write their own book.

1. Albert Einstein (1879–1955); Nobel Prize in 1921

Do we even need to justify this? He invented relativity, both special (this chapter), and general (chapters 5 and 6), virtually from whole cloth. He

---
*That's Klingon for "idiot." Please don't take our lunch money.

showed definitively that light is made of particles (chapter 2), and despite never really believing in it, was one of the founding members of quantum mechanics. His name is virtually synonymous with "genius," and—let's face it—he's the only one of the lot whom you'd recognize by sight.

2. Richard Feynman (1918–1988); Nobel Prize in 1965

Feynman had the sort of mind that makes him a hero to pretty much every young physicist. He invented the field of quantum electrodynamics, which used quantum mechanics to explain how electricity works (chapter 4), and showed that particles and fields literally travel through every possible path simultaneously (chapter 2). He also was known as "the great explainer," and at least a few of our examples in this book are stolen shamelessly (but with attribution) from the Feynman lectures.

3. Niels Bohr (1885–1962); Nobel Prize in 1922

In a little while, you're going to read chapter 2, and it's going to be all about quantum mechanics. You're going to love it! About halfway through, we're going to explain that the standard view of quantum mechanics to this day is something known as the "Copenhagen interpretation." We'll give you three guesses where Bohr was from. In addition to basically defining our modern picture of the world, Bohr also gave us our first realistic picture of the atom and showed that you can't just make an atom any old way, but that the states are "quantized."

4. P. A. M. Dirac (1902–1984); Nobel Prize in 1933

Dirac was one of those guys who plugged through a set of equations, got something that seemed physically absurd, but decided that "God used beautiful mathematics in creating the world" and assumed that the equations must be correct, anyway. This, pretty much, is how he predicted the existence of antimatter four years before it was ever detected.

5. Werner Heisenberg (1901–1976); Nobel Prize in 1932

When Heisenberg won the Nobel Prize, his citation read, "for the creation of quantum mechanics, the application of which has, inter alia, led to the discovery of the allotropic forms of hydrogen." While Heisenberg didn't exactly invent quantum mechanics, he contributed enormously to it, and invented the "Heisenberg Uncertainty Principle." More on that in chapter 2.

Quantum Weirdness

"Is Schrödinger's Cat dead or alive?"

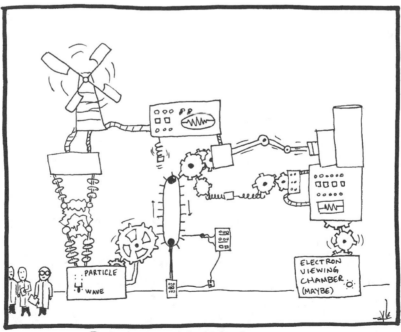

I f you're anything like us, you have contempt for authority that is matched only by your zeal for life. You take orders from no one, and sure as heck don't take anything on faith. We understand where you're coming from, being loners and rebels ourselves. This is why we won't answer your questions with "Because we said so" when explaining how the universe works. Instead, we try our darnedest to appeal to your everyday experience and your common sense, and use those to point you in the right direction.

We can't do that with quantum mechanics. If you follow your common sense, you're going to get lost, even if you don't think you are. As with Hansel and Gretel, you're likely to be drawn to the bright colors and the simple answers that come from taking the easy route out. Think of us as your breadcrumb trail, ready to lead you to the true path of quantum weirdness. Well, without the part where we get eaten by a flock of ravenous birds.

"What's so strange about quantum mechanics?" you ask, with a devil-may-care smirk. We know—you've seen it all, and nothing could faze you. So you won't mind a little pop quiz.*

---

*If you look down and you're dressed only in your underwear, chances are that you're having that dream again.

Ye Olde Tyme Classical Intuition Quizze

Please answer honestly. Even if you're so worldly that you come into this already knowing a thing or two about the quantum world, no fair trying to pretend that your intuition accepts paradoxes with ease.

QUESTIONS:

1. Do you accept Robert Frost's contention in "The Road Not Taken"?
   Two roads diverged in a yellow wood,
   And sorry I could not travel both
   And be one traveler . . .

2. Consider Hamlet's dilemma: "To be or not to be." Do you really need to choose between the two?

3. If a tree falls in a wood, does it make a sound?

ANSWERS:

If you answered yes to these questions, then congratulations! You have a mind well suited to living in the classical world.

If, on the other hand, you answered no to any of them, then you've failed the classical intuition part of the quiz, but you might be ready to step into the quantum world.

If you passed the classical intuition quiz, then you are in excellent company. Sir Isaac Newton (and his successors) helped us to build trains, cars, and even spaceships based on strong classical intuition. And unless you, personally, are designing microchips, there's a fair bet that virtually all of the interactions in your everyday life are classical as well.

But there's a lot going on beneath the hood, and if you look closely enough, the physical world is *really* ruled by the microscopic realm of quantum mechanics. We'll get into the details in a bit, but we should at least start by explaining the name. The "quantum" part describes the following phenomenon: if we're looking at the energies of electrons or any other particles, they can't just have any old value. In much the same way that you can purchase only a 40-, 60-, or 100-watt lightbulb (and not a 93-watt bulb), energies in the microscopic world can only (or rather, *should* only) come in "quantized" states. The other side of the "quantum" comes from the fact that we sometimes will talk about all of space being filled with something like, say, an electric field. However, if we look at things in detail, we'll find that it can be broken down into individual particles.

The "mechanics" part? That's just filler. To help us illustrate our points, we'll be spending some time with two individuals who epitomize the essence of quantum weirdness: Dr. Henry Jekyll and Mr. Edward Hyde. While Dr. Jekyll is kind, good-natured, and pleasantly predictable, Mr. Hyde is a fiendish beast of a man, detested in ways reserved almost exclusively for mass murderers and karaoke enthusiasts.

Of course, there's something you should know about Dr. Jekyll and Mr. Hyde: the two are not mutually exclusive. Mr. Hyde is a deformed, grotesque renegade living within Dr. Jekyll and popping to the surface to cause trouble and wreak havoc.* Whether he does it randomly, or when the mood strikes him, or at some designated time, a transformation can turn Jekyll from a mild-mannered physician into a foaming, enraged sociopath in no time flat.

To get things rolling, we join Dr. Jekyll, who is on a stroll in the freshly fallen winter snow. Enjoying the crisp December air, Jekyll comes upon a white picket fence with a plank missing. Because of his pleasant nature and joy of harmless pranks, Dr. Jekyll stands a few feet back and begins flinging snowballs. Many of the snowballs hit the fence (he's a scientist, after all, and his aim is nothing to write home about), but a few make it through the missing plank, splattering the house standing some distance beyond. The pattern, as you might expect, is pretty simple. There is a messy but quite unmistakable vertical streak of snow on the house.

Not content with this boring target, Jekyll wanders the neighborhood until he comes upon a fence with *two* planks missing, so that there are two distinct gaps in the fence. He again tosses snowball after snowball,

---

*Much like a pimple before the prom.

and *"Pow!" "Splat!"* some go through the left gap, some through the right, and some hit the fence. Looking at the wall of the house beyond, he sees two unmistakable lines of snow. We can say with pretty high confidence that the blobs of snow on the left must have gone through the left gap, and vice versa.

Dr. Jekyll's double-slit experiment is based on the design by the English physicist Thomas Young, and in this case, it clearly illustrates the epitome of particle behavior. We make one slit in the fence and we get one line of snow, and by introducing a second slit, we get a second line. We could do the same experiment with stones or custard pies and pretty much get the same result. The point here is that the outcomes of Dr. Jekyll's experiments are safe and predictable, and attune perfectly with your intuition. If a bobby* sees Dr. Jekyll chucking snowballs at a home and takes chase, we know at the beginning of the pursuit the exact location of both of them. Likewise, when Dr. Jekyll ducks into an alley, we again know exactly where he is. Since we can measure how long the city blocks are and for what amount of time he ran, presumably we know how fast he was running as well.

This is proper behavior for a particle, though perhaps not for a gentleman.

We haven't really said anything shocking. But what if we remove the rose-colored glasses of classical mechanics and look again? We're going to find the equivalent of Dr. Jekyll both turning down an alley *and* continuing down a street; of having a snowball go through *both* gaps in the fence.

Is light made of tiny particles, or a big wave?

We can only spend so long convincing you of your ability to understand the classical world before delving down into the microscopic world in which quantum mechanics holds sway. As a first step, consider a humble light beam. In the seventeenth century, Newton argued that light must be made of individual particles of light called photons. Using prisms, he separated sunlight into different colors, and argued that light must therefore be ultimately composed of little particles.

Around the same time, the Dutch physicist Christian Huygens came to precisely the opposite conclusion. He showed that if we imagined light as emanating from a single point, much like the surface of a pond after a pebble has been dropped into it, then he could explain all observed light phenomena. He claimed that light behaved like a wave.

Before you can truly appreciate why this is such a bizarre dichotomy, we need to explain what a wave is.

---

*You Yanks might know him as a police officer.

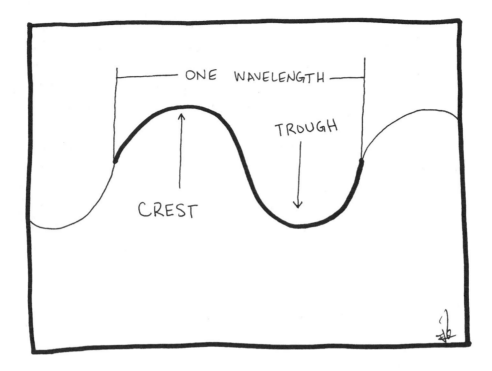

You have seen waves before at a beach or (god willing) in your bathtub. The water waves in your tub, the sound waves in the air, and light waves all have some common properties: amplitudes, speeds, and wavelengths.

The height of the crests and the depths of the troughs (also called the amplitude) tell us how strong the waves are. For you to be able to listen to Foreigner on your FM radio, the sound must first be converted to a series of crests and troughs, and beamed out from a radio transmitter. The amplitude of those radio waves control the strength of the signal, and thus how clear "Hot Blooded" will sound coming out of your stereo.

Waves also have some kind of propagation speed. Radio signals are just a specific type of light wave, and all light travels at 299,792,458 meters per second. This is not just because DJs know you need your classic rock fix, and that it's urgent.[*] After the radio wave reaches

---

[*]*So* urgent.

your antenna, it is converted into a sound wave (which is created by the motion of your speakers), which hits your face at about 340 meters per second. This means that with rare exceptions, it takes less time for the radio signal to travel from the radio station transmitter to your radio than it does for the sound wave to travel from the speaker to your ear.

Finally, there's the wavelength, which is the distance between successive crests or troughs, and the mechanism that carries all the information about the color and energy of a wave. Visible light has a wavelength a bit less than 1/1000 of a millimeter. Lower-energy waves, such as radio waves, can be many meters long. Higher-energy waves, such as X-rays, have a wavelength of 10^{-9} meter or so, and at even higher energies we have gamma rays. You don't want to mess with those, since they most frequently grant any exposed individuals with superpowers.[*]

These two pictures—the particle and the wave—seem very different. On the other hand, it turns out that under some circumstances, both predict the exact same things. When shined on a mirror, we know that light reflects off the mirror and gets absorbed by your eye.

Reflection is very easily described by the particle view of light, and by way of analogy, consider a photon as a ball. If you were anything like us, the closest you ever got to "playing catch with the guys" was tossing a tennis ball at the garage door. One flaccid lob, a loud thud, and an awkward swat later, the ball was back in your hand. If you concentrate really hard, you might remember someone saying how playing catch works: "The angle of incidence equals the angle of reflection." On the other hand, it may be that if you concentrate really hard, all you hear is the theme song to *Knight Rider*. That's fine, though; just take our word for it. You know all about the reflection of photons. If we replace

---

[*]See, for example, the Incredible Hulk. The Fantastic Four, on the other hand, got their powers from *cosmic* rays, which we'll see in the next chapter.

a tennis ball with a photon and the garage door with a mirror, then we are describing light perfectly.

Of course, waves reflect in exactly the same way. Think about the design of a violin or a concert hall. The acoustics are all determined by what will happen to a sound wave as it bounces around the room or the cavity. And just like in the particle picture, the reflection of light is given by the magic relation that "the angle of incidence equals the angle of reflection."

This whole wave/particle dispute may seem like nothing more than semantic quibbling, since the two predict the same relations for reflection. But rest assured, particles and waves don't *always* predict the same thing.

What makes a wave interesting, for our (and Huygens's) purposes, is that two waves can interfere with each other. Drop a couple of pebbles in a calm pond and see what we mean.

The physical phenomena are easy to interpret in any way you like, but they didn't settle the important question: is light comprised of electromagnetic waves, or is it made of particles? This dispute went back and forth for hundreds of years until the twentieth century, in which, much like a children's talent show, everyone was declared a winner. To see how, let's turn back to our man Jekyll.

After a long day of throwing snowballs and gently ribbing the local authorities, Dr. Jekyll returns to his home and laboratory, eager to run a few experiments. There, where he has more civilized scientific apparatuses at his disposal, he can run Young's double-slit experiment the way it was meant to be run. That means, instead of fences and snowballs, he'll use a screen with a tiny vertical slit and light shone from a laser pointer. Behind the front screen there is a rear projection screen on which we can see what pattern the light makes. What do you expect he'll see?

Don't overthink it. He'll see a single bright fringe projected on the rear screen.

On the other hand, things get a little more complex if he cuts *two* slits in the front screen.

As he does so, he finds himself turning into his beastly self: Mr. Hyde. Light goes through both, and the wave from one interferes with the wave from another, creating a complicated pattern on the screen beyond.

From Young's original notes, we can see the double-slit apparatus from above:

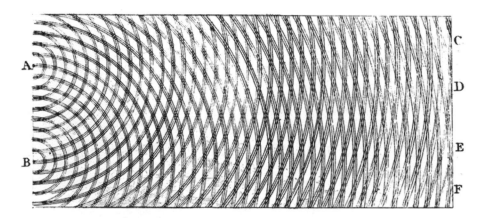

The light, coming through holes A and B, travels to the opposing screen and makes bright spots at C, D, E, and F (and at other points above and below where Young cut off his diagram). Look familiar? Like you dropped a pebble in a pond at point A and point B? This is just a more precise version of what waves interfering with each other look like.

If you get nothing else out of this discussion, you should know that the multiple bright fringes are a sure sign that wave interference is at work. Light must be passing through both the left slit and the right slit at the same time to interfere with one another; otherwise we wouldn't get the complicated pattern seen on the opposite screen.

Unlike reflection, there's no way that we can get interference from light behaving like particles. If you hold a billiard ball in each hand and crack them together, you won't get places where the balls cancel each other out. Instead, they'll just bounce off each other. Only waves can be added together to interfere.

So here's your handy guide:

- Two bright fringes = particle behavior (Jekyll).
- Many bright fringes = wave behavior (Hyde).

Can you change reality just by looking at it?

Light is clearly a wave. Young's double-slit experiment demonstrates it beyond a shadow of a doubt. Case closed, right?

Not on your life. Newton was absolutely convinced that light had a particle nature, and he wasn't the only one. In 1905 Albert Einstein showed that light is *really* made of photons. Grand pronouncements need proof, no matter how many big names in the field shout down how convinced they are, and so Einstein explained his position with the "photoelectric effect."

Scientists had observed that if you shine a beam of ultraviolet light on metals, electrons will pop out. On the other hand, exposing the same metal to less energetic wavelengths produced no effect. Einstein reasoned that the only possible explanation for this phenomenon, dubbed the photoelectric effect, was that light must be fundamentally made of individual particles, which then each transfer their energy to a single electron. It's like hitting a billiard ball with a cueball, and that sounds more like a particle than a wave, right? Since red or green or blue light (on a photon-by-photon basis) is so weak, no single photon would be energetic enough to knock out an electron; hence the observed effect was noticed only at high energies.

Though Einstein won the Nobel Prize for this discovery and though almost every introductory book on the subject gives him credit for proving that light behaves like particles, his proof is actually somewhat inconclusive. In 1969 a few research groups showed that you could explain the photoelectric effect through the wave nature of light. Einstein's explanation was successful at explaining the effect, but it turned out not to be the *only* explanation. Though his proof had a few loopholes, it turned out that he was right, anyway. Many subsequent experiments showed that light definitely behaves like a particle.

All of this debate might seem to rank up there with questions that affect your life similarly, such as "How many angels can dance on the head of a pin?" or "Where is the cast of *Blossom* now?" Who cares whether light is "really" a particle or a wave? And on the face of it, this might not even seem like too much of a contradiction. After all, while water in the ocean clearly exhibits wavelike behavior, we know that it's really composed of individual (particle-like) molecules.

Maybe light behaves in the same way. Perhaps light only seems to be a continuous wave in the same way that the picture on your TV appears continuous. If you press your face up to a TV screen, it's "really" made of individual pixels.

Does light appear like a wave only because there are so many photons involved? In the context of the double-slit experiment, perhaps a big bunch of photons go through the left slit, and a big bunch of photons go through the right slit, and then the two waves are interfering with each other.

If only life were so simple.

We mentioned before that your physical intuition wouldn't do you very much good in quantum mechanics. We hope you haven't thrown away your water wings, because you're about to get tossed into the deep end.

Lots of photons go through each slit and interfere with one another, demonstrating wave behavior. Mr. Hyde, wanting to revert to Dr. Jekyll, has an idea. "Perhaps," he growls to himself, "if I turn down the intensity of the beam, only one photon will go through at a time. And a single photon can't *possibly* behave like a wave because it will have nothing to interfere with."

Oh, the poor, deluded lummox. Let's watch what happens as he carries out his misguided project.

As planned, he turns down the beam so he knows for a fact that only one photon at a time is heading into the apparatus. As before, there is a detector on the back screen that counts every photon striking it. Even though it would take a while for the counts to accumulate, Hyde could look at the pattern that they form on the back screen.

Hyde sees a pattern of fringes on the back screen indicating that the photon beam is, indeed, exhibiting wave behavior. The photons coming in are interfering with something. But the beam is set to fire only one photon at a time. The only logical interpretation is that the photons are interfering with *themselves*. Every photon passes through both slits simultaneously. Frost was wrong. You can (at least if you're a photon) travel both paths, not just the one less traveled.

We know that the photon can behave as a particle or a wave. Knowing that the photon *can* exhibit both properties doesn't explain how it knows *when* to exhibit one or the other. In 1978, John Archibald Wheeler of Princeton University proposed an interesting experiment to see how photons interact with a double-slit experiment if we were to change the rules midway through. "Imagine," said Wheeler, "if the back screen were removable, and some distance beyond it were two little telescopes, each pointing directly at one of the two slits."

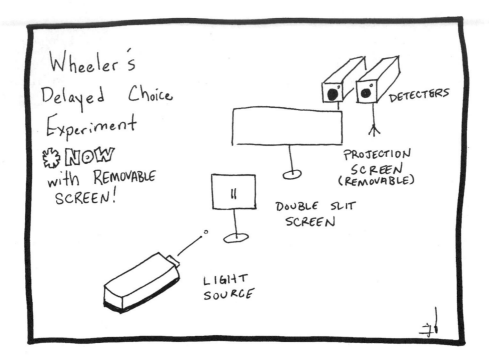

If the screen were removed, we could tell, presumably by looking through one telescope or another, which slit a particular photon went through. For that to be the case, the photons would have to go through one slit or the other, but definitely not both. In other words, we can *force* the photons to behave like particles by removing the screen, and subsequently force the experimenter from Hyde back to Jekyll. If we replace the screen, then the photons start behaving like waves again, and the dastardly Hyde reigns once more.

The fact that we can change the behavior of photons by adding or removing a screen is weird enough, but what Wheeler proposed makes it even stranger. What happens if we remove the screen *after* an individual photon passes the first screen (the one with the slits)? The "delayed choice" experiment allows us to turn the light from particle to wave or vice versa at any point in the experiment.

In other words, after it has already occurred, we can make the photon have gone through[*] only one hole simply by removing the back

---

[*]The fact that we can change the past by something we do in the present not only creates problems with our understanding of physics, but also with verb tenses.

screen. What's more, by our actions, the photon somehow chooses to go through one slit or the other. It's deeply spooky to be able to affect reality in such a profound way, especially since the photon seems to make the choice retroactively.

Before we forced the photon to behave classically (by removing the screen), quantum mechanics—and Wheeler—says that there was no way we could have predicted which hole the photon would have gone through. We can, in fact, change the quantum world *after* some event should have occurred.

We are left with a couple of staggering implications:

1. Our observations of a system fundamentally change it.
2. Individual photons can behave like particles or waves, and can switch between the two in the blink of an eye.

If you look at them closely enough, what are electrons, really?

The strangeness of quantum mechanics would be all well and good if it only corresponded to light. Light is special; after all, it's massless and, like a sailor, constantly moving at c. The problem, as you may have guessed, is that the effects of quantum mechanics seem to extend beyond photons.

Electrons are the lightest particles we can work with easily. If you don't know too much about electrons, that's fine; we talk about them quite a bit more in chapter 4. For now, all you have to know is that we work with electrons all the time. Old (nonplasma) TV sets were built around cathode-ray tubes, which is just a fancy way of saying "ballistic electron gun that fires particles at your face at sublight speed."

What happens if we shoot electrons at a double-slit experiment and then put a fluorescent screen behind it? Every time an electron hits the fluorescent screen, we get a flash of light, so we can count how many photons hit any particular part of the screen. If Hyde had been able to get his grubby, despicable hands on an electron beam, and if he turned down the source so he could send through only one electron at a time, he would *still* get the wavelike behavior on the screen. This is the same behavior we observed with the photon!

For practical reasons, this experiment couldn't be conducted until relatively recently, although within the physics community there was almost no doubt as to how it would ultimately play out. In 1989, Akira Tonomura of Gakushuin University and his collaborators performed the double-slit experiment with electrons, and you may not be the least bit surprised to learn that using an electron beam, they got exactly the same wavelike pattern of multiple fringes on the back screen (see below) as we did using a light beam. At least we *hope* you aren't surprised.

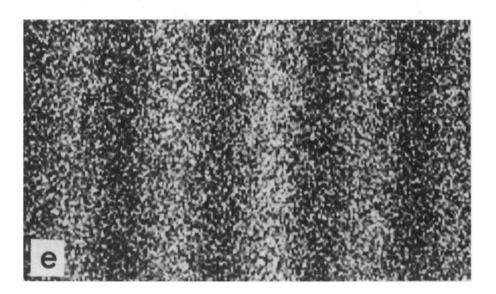

And just in case you need Hyde to beat you over the head with what this means, we'll repeat it: the fact that an electron can interfere with itself means that in reality it goes through *both* slits at the same time. But no matter how sharp your Ginsu knives are, you can't possibly split an electron in twain. How's that for a paradox? The electron goes through both slits without ever breaking in two.

Of course, this isn't just true of photons and electrons. The same experiment has been performed recently with all sorts of microscopic objects: neutrons and atoms, for example. All of them reveal this same quantum weirdness.

We admit that we're really giving you the hard sell on the double-slit experiment, but we assure you, it's absolutely necessary. Topics such as relativity allow a physicist to assume a fact such as the constant speed of light, and then use theory to figure out pretty much everything else without ever leaving the comfort of his or her parents' basement. Quantum mechanics was, quite conversely, almost entirely driven by experiment after experiment in which previous theories had failed to explain what was going on.

The other side of Tonomura's experiment is the same as Wheeler's delayed-choice experiment. If we somehow monitor the electrons to see if they are passing through one slit or the other, then we collapse the wave function, once again forcing them into particlelike behavior.

"Collapsing the wave function" is one of those phrases that physicists just toss around, along the same lines of, "computing the eigen values of the Hamiltonian" or "staying home alone on a Saturday night." We're just so *used* to it that it might never occur to us that it needs additional explanation.* But on the wave function, perhaps, a few more words might be useful.

In the quantum model, everything is a wave. Electrons, if you look at them closely enough, aren't little marbles—they more closely resemble little clouds. Where the cloud (or "wave function," if you prefer consistent terminology) is thickest, we have the highest probability of finding the electron at any given moment.

When we say that an electron "behaves like a wave," or if you hear talk about an electron cloud, this doesn't mean that the electron itself is really an extended, flimsy object, like cotton candy. We also don't want you to think of an electron's wave function like the Tasmanian Devil from the old *Looney Tunes* cartoons—something moving so fast that it's simply a blur.

The electron really *is* in many places at once, and by measuring its actual position, we change the nature of the system. We have no way

---

*Not that "staying home alone on a Saturday night" really needs much additional explanation. Enjoy the rest of the book, nerd.

of knowing ahead of time where the electron really is, and it's only by observing it that we isolate it. At the instant when we measure the position of an electron (by hitting it with a photon, for example), the wave function collapses, and for an instant afterward we know almost *exactly* where the electron is. The wave function no longer extends over a large region of space.

Imagine Jekyll and Hyde sitting down to play a game of Battleship.[*] Hyde, as we know, is a cheater, and so for a while, as Dr. Jekyll calls out coordinates, Hyde keeps announcing miss after miss while he moves his ships around. Eventually Hyde realizes that the charade can't go on forever, so he's forced to put his battleships somewhere on the board and declare a hit. Jekyll's measurement of the boat's position clearly affected it.

To put it another way, think back to your youth. When you were young, the whole world was your oyster. Options abounded: did you want to be a nuclear physicist? A cosmologist? An astronomer? Now consider what you've accomplished. All of that potential and uncertainty collapsed into a single state of what you've *actually* done with your life—one path.

Is there some way I can blame quantum mechanics for all those times I lose things?

Having introduced the basic idea of quantum weirdness, we're going to spend a few moments talking about some seemingly impossible repercussions—the ones where you're most likely to think there's some sort of trick or oversimplification.

When we shine a beam of electrons at our double-slit experiment, we don't know which slit a particle goes through. This is another way of saying that there's uncertainty in the position of an electron. In 1948, Richard Feynman, then at Cornell, realized something even odder about this experiment.

---

[*] Yes, we're aware that they're technically the same person. This is what we call an analogy.

To visualize exactly what Feynman did, let's set up the experiment again. Hyde shoots his electron beam at the double slits and sees the results. "What if," he wondered, "we were to cut a third slit in the front screen?" Being the murderous type, Hyde produces a blade and slices another slit in the screen. Now the electron's wave must go through all three slits, each with some probability, and all three subsequent waves would interfere with one another.

"And a fourth and a fifth?" Again, the electron will pass through all of the slits simultaneously. "And what if we keep cutting out slits until the entire screen is gone?" Hyde tears into the sheet like it was made of English street urchins, until the whole thing lies in tattered shreds on the laboratory floor. The electron has to pass through everywhere that the screen originally was, with some probability.

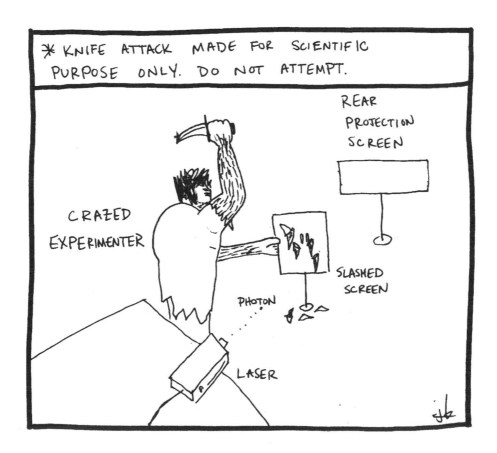

What would happen if Hyde put many such (empty) screens between the beam and the rear projection screen? Naturally, the electron would pass through all of the slits in all of these screens with a probability given by the wave function.

But if there were no screens at all, then what Feynman is describing is a situation in which an ordinary particle is simply moving from point A to point B, and in case you've missed the point (which admittedly is a little subtle), what he's effectively shown is that in going from one place to another, particles don't necessarily travel in a straight line, or take a curved path, or take any other particular route. They take *all possible* routes, each with some probability!

Stranger still, while taking all of those possible routes, the particles do all sorts of impossible things. They have the "wrong" mass, for example, or appear to travel faster than the speed of light. What would normally seem impossible simply happens with a very, very low probability. But still, those "impossible" eventualities need to be included in the calculation to get it right.

We know, this sounds not unlike the chemical-induced "philosophical" conversations you had with your friends late at night in college:

"Hey, man, what if we're, like, everywhere at once?"

"Whoooooa!"

But understand that much like the double-slit experiment, Feynman's "all possible paths" description is a useful picture of reality because it gives us the right answer. Since we never measure the particles between the front and the rear screens, we can't be sure of where they are anywhere in between. And if we were to measure their position, we would disrupt the system.

The idea of never knowing exactly where the particle is without changing something must seem a little frustrating. We agree. Still, this thought experiment is helpful in imagining the nature of moving particles, even if it does hurt your brain.

That said, if you've misplaced your car keys, don't look to quantum mechanics to help you. Quantum mechanics deals only with the probability of finding a particle in one place or another, but that doesn't mean it's vague about the details. It's very, very precise about how little we know about the universe.

In 1927, Werner Heisenberg, then at Göttingen, postulated that for any particle, not only couldn't we know things such as the position or state of motion of a particle, but that the better we know the position, the worse we will be able to measure the velocity,* and vice versa. As a result, if we knew the position of a particle to infinite precision, we would have no idea what the speed of the particle is. Likewise, if we somehow knew exactly how fast a particle was traveling, we would have no idea *where* the particle was.

The Heisenberg Uncertainty Principle is one of the most misunderstood concepts in quantum mechanics, mostly because people tend to assume that it's really just a classical phenomenon. Many introductory books on quantum mechanics erroneously "prove" the Uncertainty Principle with the following argument. If we want to figure out where a particle is, we have to hit it with a photon. If the photon has a very long wavelength, then we aren't able to measure the position very accurately. Long-wavelength photons don't pack much of a punch, so the electron isn't affected very much by the measurement, and we're able to measure the speed very accurately.

At the other extreme, to get a very good idea of where the particle is, we need to hit it with a short-wavelength photon. A short-wavelength photon is extremely energetic, which means that it will give the particle a big kick. As a result, we won't know the subsequent velocity of the particle very well.

From that, you might guess that the photon is what makes the position and speed of the particle uncertain. After all, without the photon hitting the particle you're trying to observe, you wouldn't mess it up. This isn't actually the case, though. While our observations (the introduction of the photon) affect the state of the particle, the uncertainty of its location and velocity is *fundamental*. There is no getting around it.

The Uncertainty Principle has some surprising results. Let's first consider Dr. Jekyll in his laboratory, moving notebooks onto his lab bench. If he leaves for tea and returns to retrieve the notebooks, they

---

*Technically, the momentum. If you know enough about physics already to distinguish between velocity and momentum, then you can stay after class and clean the erasers.

will be exactly where he left them—they are big, weighty things and not likely to move by their own volition.

But what if we assume that Mr. Hyde comes out to play? In his infinite cruelty, Hyde has ignored the notebooks entirely and instead confined an electron to an incredibly small box.* Knowing that the electron is in the box, we realize it must have a very small uncertainty in position. By virtue of that, it must have a relatively large uncertainty in its velocity. What do we mean by "uncertainty"? We mean that nobody knows or *can* know what the speed of the electron is. Hyde does know that it's not sitting still, though. If it were, he could say in a confident voice that the velocity is zero. It *must* be jostling around inside the little box.

Perhaps it's going quickly to the left, and perhaps it's going quickly to the right. He simply can't know. The smaller he makes the box, the better his knowledge of the electron's position, the worse his knowledge of its velocity, and consequently the faster the electron may be jostling around.

But it doesn't end there. Uncertainty isn't just for electrons. As we've already seen, light is made of waves as well, and as we'll see in the next chapter, light is just one of four (or perhaps five) fundamental fields permeating our universe. What happens if Hyde takes a small "empty" box completely devoid of electrons or light?

We've already said that Hyde is a madman, and it turns out that his little experiment is completely impossible. No matter how hard he tries, it turns out that light is always going to figure out a way to make it into his box. To understand this, you first need to realize that even though he doesn't put any light in his box, in principle, lots of individual light waves *could* fit into the box. Like the electron, the amplitude of these waves is uncertain, but Hyde is trying to set those amplitudes to zero. This is the foundation of Quantum Field Theory and represents the marriage of special relativity (chapter 1) and quantum mechanics.

In exactly the same way that confining an electron to a small box makes it bounce around with greater and greater average energy, uncertainty

---

* Truly, the man is a monster.

guarantees that there is no way to make an electric field disappear entirely.

This means that even within Hyde's supposedly empty box, photons pop into and out of existence constantly. This is crazy (ah, but so is he). It means that even empty space has energy in it. This is known as the vacuum energy of the universe, and it has some very weird properties. For one thing, if Hyde were to smash the box like an accordion, even though the volume would get smaller, the density of the vacuum energy wouldn't go up. This is very different from just about everything else.

Normally at this point in the conversation a nonphysicist will accuse us of just "making stuff up." After all, if the universe is filled with the vacuum energy, why don't we see it? It accounts for an awful lot of energy, after all.

Perhaps this will make more sense in the context of helping a friend move.* Say he lives in a fifth-story walk-up. Volunteering seemed like a good idea at the time, but now you have to carry his dresser up four flights of twisty, turny stairs. By the end of the day you can't help but notice how much work it is to climb to the fifth floor. But did it ever occur to you to notice that he lives two thousand feet above sea level? Why would it? It never comes into play. Likewise, the vacuum energy is like the ground floor of the apartment. It's the lowest energy you'll ever measure, and everything else will be measured in comparison to it. In much the same way, vacuum energy is one of those things you never measure "less than."

This still doesn't prove that we're not just making stuff up. We only showed why we never notice that the vacuum energy is around, but we haven't actually given any good reason to believe that it's real in the first place. That's a story that will have to wait until we talk about the nature of space. Vacuum energy, for now, is a repercussion of quantum mechanics and, like Hyde, a necessary evil.

Of course, since the vacuum supplies a ready amount of energy, it means that by way of $E = mc^2$, the universe can constantly make particles. Much like a boiling pot of water, a particular can bubble into the vacuum, with

---

*Or in the case of Hyde, helping a fiend move.

the caveat that it doesn't last very long. Particles can be created, but they very quickly annihilate, and the more massive the particle, the shorter they live before disappearing forever.

Can I build a transporter, like on *Star Trek*?

We're not used to thinking of things like electrons as wave functions, but wave functions they are. This means that to a greater or lesser extent (mostly a lesser one), the probability of detecting an electron extends over great distances—technically over the entire universe. Things that we've come to think of as impossible should really be redefined as just incredibly unlikely.

Imagine that the people of London laid a trap by building an enormous hole in the ground and that Hyde fell in. Mr. Hyde could try to jump out, but his stout legs are not enough to get him out of the hole. In the parlance of physics, we'd say that he doesn't have enough energy to escape. But wouldn't you know it? Quantum mechanics dictates that because the position of the criminal mastermind is uncertain, there is a possibility that Hyde is "observed" outside the hole. That's just a fancy way of saying that he gets out. A professional escape artist, he tunnels out of a position that one might have thought was impossible. He does not tunnel in the classical sense—that is, through the dirt, with a teaspoon—he just appears outside the hole.

Let's make one thing clear. Hyde can't control this tunneling; it's just something random that happens occasionally. What's more, for a big object such as our maniacal friend, we'll be waiting a very long time before anything happens—probably much, much longer than the age of the universe.

For microscopic objects such as atoms, on the other hand, tunneling is not only possible, it's almost inevitable. Uranium, plutonium, or thorium can sit around happily for a while, with all of their constituent particles safely inside their nuclei. Think of uranium as being "made of" a nucleus of thorium plus a nucleus of helium. The two of these are bound together so tightly that it seems impossible (to the classical mind) that the helium (the lighter of the two) could ever escape. But just wait! As unlikely as it seems, after about 4.5 billion years, there's a decent chance that the helium can tunnel out and escape.

Not only does quantum mechanics afford us the opportunity to become the ultimate escape artist, we'll also throw in, at no extra charge, teleportation! Since the wave function of an electron, uranium—heck, even Mr. Hyde—technically extends over the entire universe, there is a non-zero* chance that you or anything else could suddenly be observed in some other star system.

This isn't what you wanted, we know. You wanted a "real" teleportation device, such as the kind you saw on *Star Trek*.† You wanted something where you could really control where and when your away team was sent, not something that happened by pure chance. Well, you're in luck. Quantum mechanics can help you build a bona fide teleportation device, but before you send in your cereal proofs of purchase to buy one, we need to give you a few caveats about how they actually work.

First, a realistic teleporter doesn't actually move your atoms from point A to point B. Instead, it makes a perfect replica. Suppose you wanted to send a statue across the room, since you're too nervous to try this on human subjects. The receiver would have to have a bunch of carbon atoms, a bunch of iron atoms, a bunch of calcium atoms, and so on, all at the ready. The transmitter would have to send a signal that gives the receiver precise instructions describing the wave function of every atom and the overall configuration of the statue. If the wave functions are exactly copied at the receiving end, then what we have really is teleportation.

This doesn't seem quite right, since we only *copied* the statue, we didn't actually move it. Let us ask you something. What's the difference? The copied statue would look exactly the same, down to the minutest detail. It would weigh the same, feel the same, and so on.

As far as the laws of physics are concerned, the statue would *be* the same. The universe doesn't distinguish between one calcium atom (for instance) and another. They're all identical. What's more, the process of sending the signal to the receiver destroys the wave function of the

---

*A physicist's way of saying "vanishingly small." We also use the term "nontrivial" to mean "nearly impossible."
†Admit it, Poindexter. You own your own unitard with a Starfleet insignia.

original. In other words, our teleportation device isn't just a fax machine, since you start with one and you end with one—only in another place.

So much for teleporting statues. What happens if we teleport a person—you, for instance? The teleported version of "you" would never know the difference. What are "you" besides the sum of the wave functions of all of your quadrillions of atoms? Those atoms encode not just your appearance, but your memories as well. And since the original from which you'd been copied had been destroyed, there's no other "you" to contest your story.

Sounds too good (or spooky) to be true? It isn't, but first we have to clear up a little detail. Throughout this chapter, we've been talking about the wave function of individual atoms. That said, in reality if two atoms interact with each other, it's more appropriate to describe the combined wave function of the *two* atoms. The atoms are said to be in a state of "quantum entanglement," which is simply a complicated way of saying that if we know something at the quantum level about one atom, then we know something about the other.

Here's the basic procedure:

1. Take two atoms (A and B) and entangle them,[*] giving A to the transmitting end of your teleporter and B to the receiver.
2. The transmitter takes a *different* atom, the one he'd like to teleport (call it C), and interferes it with A. In the process, the wave function of A collapses, and so does B at the receiving end. We've seen that interference and observation have this effect on wave functions before, and as a result, C gets changed as well. This is equivalent to saying that your transmitted object gets destroyed.
3. The receiver does the same thing on his end, but interferes the target atom, D, with his own, changed, and entangled atom, B. His interference also affects D, but has the opposite effect, and D acquires the original wave function of C.

Teleportation is incredibly difficult. It wasn't until 1997 when people were first able to teleport a single photon, and until 2004 when several groups teleported an individual atom, and even then it was only a distance of a few meters. Given the amount of work involved, it would be easier for you to simply carry the atom from one place to another.

---

[*]We'll tell you how in the next chapter.

The larger the system, the more complicated it gets. Even teleporting a single molecule is well beyond anything we can do experimentally. So while teleportation is technically possible, it's going to be a very, very long time before teleporting a human being is even remotely possible, and even then, we wouldn't recommend it.

If a tree falls in the forest and no one hears it, does it make a sound?

Our examples have centered on microscopic particles, but there's nothing particular in our reasoning that says a particle has to be tiny to behave quantum mechanically. In fact, we've been arguing that our universe is fundamentally quantum. Come to think of it, if the quantum rules hold sway on the microscopic scale, shouldn't *we* be governed by the rules of quantum mechanics as well?

Yes and no.

Take the Uncertainty Principle.* We skimmed over a bit of math (read: all of the math) when we talked about it, so we'll add one more detail now. The more massive a particle is, the more accurately we're able to measure both its position and its speed.

For instance, imagine that we do the double-slit experiment with a stream of electrons. If the two slits are separated by a millimeter, we can suppose that the uncertainty in the electron's position is about a millimeter. It must be, since we aren't sure which slit the electron went through. Crunching the numbers, we'd find that the speed of the electron is uncertain to about a tenth of a mile per hour. Not a huge number, but a measurable one.

What if we measured the speed of Hyde (say, as he was running from the scene of a murder) accurate to within a tenth of a mile per hour? This is far more accurate than any device you'd be likely to carry around. Presumably, since we only measure Hyde's speed to some measurable degree, there must be uncertainty in his position as well. And there is. Hyde's position is uncertain to about one tenth of a quintillionth the size of the nucleus of an atom. On any scales smaller than that, Hyde would

_____
*Please! (We invite you to insert your own rimshot sound effect here.)

exhibit wave behavior. Since Hyde himself is much larger than that, under any reasonable situation, he's going to behave like a particle. That is, there's no *conceivable* circumstance in which macroscopic objects (like you and me, and Jekyll and Hyde) will behave like quantum objects.

Going back to the opening question for this chapter, we'll address a classic thought experiment, one that has worked its way into the public consciousness—the idea of Erwin Schrödinger and his eponymous cat.

Let's have Hyde, that merciless ruffian, construct a box with a vial of poison in it. If a particular radioactive atom, also inside the box, decays in a certain amount of time, the poison will be released into the box. If the atom doesn't decay, no poison will be released. He then places a cat into the box and closes the lid.*

After the requisite amount of time passed, is the cat dead or alive?

RECENT EXPERIMENTS IN QUANTUM MECHANICS

---

*Just to be clear, Schrödinger never actually did the experiment. It does, however, paint his psyche in a disturbing light.

This question, originally put forth as an aside in a longer technical paper by Schrödinger in 1935, didn't take up too much more ink than we do here. And though the riddle of Schrödinger's Cat doesn't tell us anything new about building a quantum computer or a microchip, it does provoke questions about the true nature of our universe. It turns out that there's more than one way to poison a cat—or at least how to interpret the poisoning.

1. The Copenhagen Interpretation

In 1927 two of the founders of quantum mechanics, Niels Bohr and Werner Heisenberg, formulated an early version of what has come to be known as the Copenhagen interpretation of quantum mechanics. This is basically what we've been implicitly assuming throughout:

1. A system is described completely by its wave function.
2. The wave function indicates that certain measurements are only probabilistic.
3. Once we make a measurement, the wave function collapses, giving us a single number.

And though we're going to describe a couple of other ways of looking at things, for the workaday physicist, the Copenhagen interpretation is pretty much the consensus opinion, mostly because it allows us to do our calculations without giving too much thought as to what it all really means.*

However, even *within* the community of adherents there is some dispute as to what the Copenhagen interpretation actually says. Is the wave function a real thing? Or is it true that the only realities of a system are those things we actually observe? Personally, these questions seem like quibbling. We are much more partial to the version by David Mermin, "If I were forced to sum up in one sentence what the Copenhagen interpretation says to me, it would be 'Shut up and calculate!'"

More to the point, how is it that our observations of something actually *make* it collapse? At the end of the day, we are also made of subatomic

---

*This also allows us to be lazy, which we like, too.

particles, which should also be subject to the laws of quantum mechanics. How does the universe know how to go from a state of indefiniteness before a measurement is made, to definiteness after?

There's an even worse consequence of the collapse of the wave function. Remember when we said that your wave function extends all the way to other star systems and that it's technically possible for you to teleport there instantaneously? Well, when you are observed here on Earth, your wave function collapses, which means that your wave function elsewhere disappears. If this doesn't trouble you, it should. Something here seems to affect something light-years away instantaneously—that is, faster than the speed of light.

Let's forget about all that and simply see what Bohr tells us about the cat. Is Schrödinger's Cat dead or alive? The Copenhagen interpretation answers, "Yes."

Seriously.

"Both," it says, "each with some probability. If we open the box, we'll collapse the wave function, and one possibility will be observed."

This is absurd! It's crazy to think that the cat can be both dead *and* alive. This is precisely Schrödinger's point.[*]

We consider an old riddle in quantum mechanical terms: If a tree falls in the forest and no one hears it, does it make a sound? "No," says the Copenhagen interpretation. "In fact, a case could be made that it hasn't even fallen until there's some observable evidence of it having done so." It seems ridiculous to imagine that something as big as a tree should be so influenced by whether it's observed. True. But what's the big difference between a tree and a cat?[†] Or a cat and a nucleus?

While not all believers in the Copenhagen interpretation would agree, Bohr thought there was something special about a conscious person making the observation. If instead of Schrödinger's Cat we had Schrödinger's graduate student, there would be little doubt that the grad student, being (largely) sentient, would be in a position to observe the system himself. Why should human observation be so important?

---

[*]By creating the cat-in-the-box experiment, he was implicitly *making fun* of the Copenhagen interpretation.

[†]A: One has a bark, while the other has a meow.

Philosophically, the biggest problem with the Copenhagen interpretation can be summed up with the question, Is there a difference between what the scientist knows and what the universe knows?

Our common sense says that in the case of Schrödinger's Cat, the difference is significant. Clearly the universe must "know" whether the cat is dead or alive, even if the scientist doesn't. In some sense, the Copenhagen interpretation says that it doesn't matter whether the universe knows if the cat was alive or dead prior to opening the box. It won't change anything observable.

There's something missing here. On the one hand, we've already seen from the two-slit experiment that a direct or an indirect detection of an electron can force it from a state of indeterminacy to particlelike behavior. If we don't disturb the electron by actually looking at it, it literally goes through both slits. It's only when we have the audacity to look at it that it "chooses" just one.

If that's the case, then, how is Schrödinger's Cat so different? It's just a more complicated system, one that happens to include not just a single electron, but also the radioactive sample, the flask of poison, and the quadrillions of atoms in the cat itself. For those of us who subscribe to a mechanistic view of the universe, this leads to an untenable state of affairs because it means we have to look at the big picture.

Since all the particles in the universe are (to a greater or lesser extent) interacting, the entire universe, including scientists and the equipment, is just one giant wave function. Looking at this statement literally becomes very spooky. It means that all observations, sensations, and actions are themselves combinations of more than one possibility—albeit with one case vastly more likely than the others.

Personally, we find the possibility of a dreamlike "superposition" universe so unpleasant that we'd rather live in a universe where consciousness shapes reality.*

2. The Causal Interpretation, or You Dropped a Bohm on Me

If you are troubled by the Copenhagen interpretation (and who could blame you?), don't worry. It's not the only game in town. There are

---

*We hope you solipsists out there is paying attention.

other interpretations of quantum mechanics. They all use the same equations, or at least produce the same results.[*] However, they provide *very* different interpretations of what is really going on. In other words, we can't generally tell which interpretation is correct through experiment; we're definitely into the realm of philosophy here.

In 1952, David Bohm, then at the University of São Paulo, came up with a "causal interpretation" of quantum mechanics. Bohm disagreed deeply with the "half-dead/half-alive" answer to the Schrödinger's Cat puzzle. He thought that all of these things we've been talking about as indefinite—position, velocity, the life signs of our cat—are completely definite. But (and this is a big but[†]), even though the particle and the universe know about these definite values, that doesn't mean that *you* know them.

Bohm proposed that there must be "hidden variables" besides the wave function, and he wasn't alone. Einstein remained deeply upset by the implications of quantum mechanics and was an early proponent of hidden variables.

In Bohm's picture, the hidden variables include quantities such as position and velocity that ordinary quantum mechanics says are completely undetermined. Think of it like a jet ski on a choppy ocean. At any given moment, the jet ski is moving along with definite position and velocity. However, if you were to try to measure the position of the jet ski, it would appear to jump around in a very haphazard fashion. Likewise, in the causal interpretation, the wave function "drives" the particle, giving it little nudges so that if we were to run a two-slit experiment, the path of the electron would produce seemingly random wavelike patterns.

The causal interpretation is extremely satisfying in one regard. It tells us that there is an absolute reality, even if we aren't necessarily sure of what it is from one moment to the next. An electron really is in one place or another. There is no Mr. Hyde at all! It's just Dr. Jekyll in disguise.[‡]

What's more, it gets around a very important problem that plagues the Copenhagen interpretation. According to Bohm, there is no "collapse

---

[*]At the risk of sounding glib, we welcome you to double-check if you don't believe us.
[†]Though we can't be positive Sir Mix-a-Lot would like it.
[‡]Note to writers of *Scooby-Doo*: this would make an excellent idea for a script.

of the wave function." The wave function never collapses because all that happens when we do a measurement is that we find out where the particle was all along. By observing it, though, we do affect it, but in a manner perfectly consistent with our classical intuition.

We mentioned that the causal interpretation produces the same results as ordinary quantum mechanics. This is both a plus and a minus. Like the Copenhagen interpretation, Bohm's causal interpretation requires that it should be possible (albeit unlikely) for signals to be sent at faster than the speed of light.

And while under normal circumstances, Bohm's version works the same as classical quantum mechanics, we need to make at least one caveat. Everything we've talked about so far has assumed that low energies and particles have been around for a while. There are lots of situations where this description just isn't good enough, and we need to deal with the questions of where particles come from and what happens when things go nearly at the speed of light. Ordinary quantum mechanics has been extended to deal with these questions, but Bohm's version hasn't, which means that it can't do important things like describe how particles are made. Can it be? Only time will tell.

Let's not have a discussion about what can and can't be done with equations, because it's distracting us from our possible feline mortality. What of the cat? Is it dead or alive by this interpretation?

Bohm basically tells us that he doesn't know, but the cat *must* be either dead or alive, one or the other. We haven't opened the box yet, but as soon as we do, we'll have our answer.

What a boring answer! "I don't know. Let's check." Boring, perhaps, but it is far less brain-bending than saying "Both."

3. The Many Worlds Interpretation

There is something deeply unsatisfying about the idea that the universe could have turned out in one way or another, but somehow, arbitrarily, chose a particular path. In 1957, Hugh Everett (then at the Pentagon) suggested the Many Worlds interpretation.

Everett posited that every random event—whether an electron went through one slit or the other, for example—gives rise to two different but parallel universes, indistinguishable except for the fact that in one, the electron goes through slit A, and the other (maybe the one we live in), the electron goes through slit B. As time goes on, the universes split again and again, nearly countless times, producing huge numbers of parallel universes.

According to Everett, these worlds can then interfere with one another. Mathematically, this looks nearly the same as normal quantum mechanics. For example, if we think about an electron in the two-slit experiment, then in our universe, say, the electron might travel through the left slit, while in other universes, it might travel through the right. The wave functions of the different universes then interfere with each other, and if we repeat the experiment with many electrons, we get the many-fringe pattern that we saw above.

In this case, too, there is no Hyde. It's just that since every universe has a Dr. Jekyll performing the same experiment, these multiple Jekylls interfere with one another.

It's not just particles that split off. You do as well. If you think about yourself ten minutes in the future, the "you" corresponds to a multitude of different "yous." Which "you" do you end up as? All of them. It's just that any given "you" only remembers the history that happened in his or her universe. That means that somewhere, there's a "you" working as a film actor and a "you" designing spaceships.* Not all possibilities are equally likely, however.

At the expense of producing infinite universes, Everett also would be able to give a comforting answer to our question about Schrödinger's Cat. Like Bohm, he would say, "I don't know. The cat's either dead or alive, and the only way to find out is by opening the box. Opening the box, however, only informs us. It doesn't change reality at all."

This is basically the same answer that we gave for the causal interpretation of the universe, but there's an important twist. If the cat turns out to be alive, that's only true for our universe. There are infinitely many other universes in which the cat is dead.

Reality, it turns out, is a very local affair.

---

*This really works both ways. It means there's also a chance that you could be talking on your cell phone in a movie theater or taking a fistful from the "take a penny, leave a penny" bowl. This is why we can't look you in the eye anymore.

Randomness

"Does God play dice with the universe?"

THE BLOMBERGS

FROM LEFT TO RIGHT: DAVE; AUNT MAVIS; COUSIN HERMANN; JEFF; UNCLE LOUIE; NEPHEW BRIAN

Say what you will about the physics of yesteryear. Maybe it was boring, and you had to remember all the rules of levers, pulleys, pendulums, and the like. But at least you knew where you stood. When the twentieth century came along, all of that certainty went out the window. But if quantum mechanics just messes with the microscopic level, on our giant, human scales it seems like we can take a page from the book of Alfred E. Neuman. What, me worry?

Lots of people buy into the idea of a deterministic universe. And who could blame them? Everything we see around us can, for the most part, be intuited or accurately predicted mathematically, and as for the rest, we just assume that it's too complicated to figure out . . . for now. Albert Einstein was convinced that underlying apparent randomness there were deterministic rules controlling everything, making everything predictable. If you understand how things are set up to begin with, the laws of physics will tell you exactly how things will end up. The determinism of the universe seems built into the equations. But "seems" are for stockings, and that apparent determinism is a lie.

At a fundamental level, the universe isn't just complex, it's inextricably random. Radioactive decay, the motions of atoms, and the outcome of physics experiments all bend to the whim of some unpredictability. At its core, the universe is Einstein's worst nightmare. Randomness may be *your* worst nightmare, too. Human beings simply aren't wired very well for statistical thinking. If the odds are really obvious, or if our personal survival is at stake, our brains might give us a clue. "Don't poke that Tyrannosaurus rex,"* your brain might say. "It hasn't turned out well for other people who tried that, and odds are it won't turn out well for you."

On the other hand, go to Vegas and ask a blackjack player who has just lost ten hands in a row what the odds are of his winning the next one. He'll say that he's due for a hot streak, or that the deck has gone cold. Optimist or pessimist, he's wrong. His odds of winning his next hand are the same as they were on the last: about 50–50.

Because (we hope) you don't spend most of your waking hours in a casino, it might help if we introduce some of the nuances of randomness in a more familiar context. We'd like to introduce you to our family, the Blombergs, in the midst of a family reunion. Besides the normal demands for grandchildren, most of our irritation comes from relatives who really should know better, but for some reason persist in refusing to believe in the power of randomness.

Consider our cousin Hermann. He's a smart guy, capable of building receivers to pick up the transmissions from alien spacecraft. He also thinks that the government, scientists, and *especially* government scientists are manipulating scientific data as part of a massive conspiracy.[†] Hermann is obsessed with global warming, which wouldn't be so frustrating if he didn't believe it was made up. Lest there be any confusion on the subject: there is virtual unanimity in the scientific community that global warming is real and that it is man-made. What complicates it from a PR perspective is that according to the general scientific consensus,[‡] the average temperature on Earth will only rise about a tenth of a degree Celsius in the next ten years. That may not seem like a lot, but over time, the environmental impact will be devastating.

---

*If the Answers in Genesis Creation Museum is correct, this scenario may have actually played out for some cavemen. And who are we to doubt them?

[†]The Trilateral Commission is involved in this somehow, too. They have to be.

[‡]The Intergovernmental Panel on Climate Change, in this case. Sounds official. We're sure Hermann would take that as suspect.

Hermann lives in Philadelphia where, according to Wikipedia, the average temperature during December is about thirty-six degrees Fahrenheit. But lo, one particularly balmy Christmas, we have temperatures in the low fifties. Whenever that happens, Hermann will take a break from writing angry letters to the government to graciously concede the point; global warming sure *seems* real from his perspective. In this case, we don't want this sort of one-time measurement to make Hermann an ally. Here's why.

Sometimes the temperature is going to be above the average and sometimes below. If this range of values is large, then we won't notice small changes from year to year. It's actually not that unusual to have a temperature fifteen degrees above the average, but neither is it unusual for it to be about fifteen degrees below the average. What happens next year when we have a *cold* winter in Philadelphia, and the temperatures in December are consistently in the twenties? Cousin Hermann assumes that all the fuss about global warming was for nothing, and goes back to building his tinfoil hat. He doesn't see the problem because he focuses on the individual days and not the general trend.

Admit it. You've been guilty of worse.

Even without focusing on the futuristic hellscape that awaits our planet, Hermann still has plenty to worry about. Why do all the little particles in his glass of water move around all the time? How long will his pet neutron last? Maybe these aren't things you've ever worried about before, but each is a consequence of randomness just doing its thing.

If the physical world is so unpredictable, why doesn't it always seem that way?

At the far end of the family tree (and from the shallow end of some cloudy gene pool) is Uncle Louie. He's sweet enough in his own way: he tells bawdy jokes and is constantly asking little kids to pull his finger. Uncle Louie's nieces and nephews have paid for college with quarters he pulled from their ears. He is, however, a degenerate gambler. Uncle Louie will bet on just about anything: movie endings, hermit crab races, you name it. So while trying to evade Aunt Mavis, Uncle Louie and Dave hide in the rec room and play a friendly game of coin flipping. Provided the coin is fair, what's the harm?

To understand this game, we need to say what we mean by a "fair coin." If the coin were flipped a million times, then it would come up heads *approximately* half the time. The longer it's flipped, the more likely it is to come up close to 50% heads. The other thing that makes the coin "fair" is that each flip is independent from the last. It doesn't matter whether the coin just landed tails or heads; the *next* flip is equally likely to be heads or tails.

Now here's the rub. Even though we expect that after a million flips Uncle Louie and Dave will be close to even, we just mean that in a *fractional* sense. In real dollars, it's a different story. After a million flips, it's actually pretty likely that either Dave or Uncle Louie will have won about an extra thousand times more than half and so be up (or down) by a couple grand. If you're wondering where this "extra thousand" came from, we recommend visiting "Uncle Dave's Technical Corner." If you're not, don't worry. It's not required reading.

Uncle Dave's Technical Corner: A Bit on Statistics

We made you a promise early on to keep equations to an absolute minimum. We've been skirting the "no equations" rule for a while now, and in a math-heavy chapter such as this one, there are a few masochists who undoubtedly demand more. "Where do these numbers come from?" we hear you cry. So here's a wee bit more.

When Uncle Louie flips a fair coin, we mentioned that we could guess with fairly high likelihood that it will come up heads nearly half the time. How near? Here's a handy rule of thumb: the *range* of outcomes will be something like the square root of twice the expected number of heads (the "winning" outcome). We're cheating a little for simplicity, but it doesn't change the basic picture. When you flip a coin a million times, you'll very likely flip heads half a million times, plus or minus an extra thousand flips.

If Louie and Dave flip a million times, Louie may win a lot of money, or he may lose a lot of money, but at the end of the day, he can comfort himself with the satisfaction that he won very nearly 50% of the time. If he wins 501,000 times (a thousand more wins than half), he'll still have won only 50.1% of the time. This is truly a waste of time, and a lousy way to make (or lose) a grand.

The upshot is that just about any outcome can occur with some predictable probability. For example, in the million-flip game, Louie (or Dave) could expect outcomes with the following probabilities:

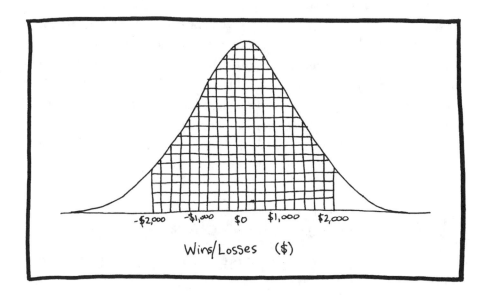

The higher the curve, the more likely the outcome. The single most likely result is that they would be even, but Louie might win (or lose) an extra grand or two and not be terribly surprised by his (mis)fortune. Those tails at either end of the graph show that it becomes increasingly unlikely that either Dave or Uncle Louie would win a huge majority of the time. Technically, Louie could flip a million times and the coin could come up heads every time. The odds of that happening, on the other hand, make the word "infinitesimal" seem positively grand by comparison.

A mathematician looking at Louie's (or Dave's) (mis)fortunes over the course of their game would describe the progress as a "random walk." To create your own random walk, find a fixed point—a lamppost, perhaps.

Stand next to the lamppost and flip a coin. When it comes up heads, take one step to the east; when it comes up tails, take one step to the west. As time goes on, you'll be equally likely to be to the east as to the west, but you will also, on average, get farther and farther away from the lamppost.*

Rather than just describe Uncle Louie's fortunes, we'll prove the point to you by flipping a million coins. How'd Louie do?

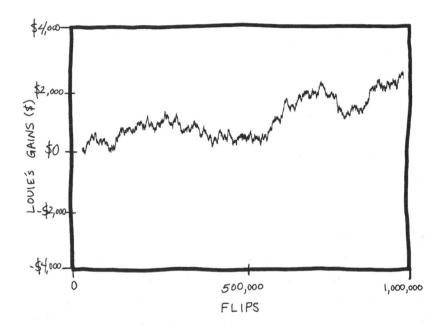

Look at that! He ends up winning about two thousand dollars off his poor, scholarly nephew, at least this time around. If you look at the rise and fall of Louie's fortune, you're bound to see trends. Maybe Louie seems to be on a winning streak for a while, about halfway through. This is not the result of a weighted coin or Louie trying to pull a fast one. It's simply the result of your brain trying to make patterns where none exists. If you've ever invested, you've probably seen the same sort of patterns in the Dow Jones Industrials average for a couple of

---

*You also run a fair risk of standing in the middle of the street, flipping coins in the air. It's best to leave the dangerous experimentation to the mathematicians.

months.* The lesson here is that you shouldn't try to time the market by predicting the random ups and downs. As our wise uncle Mortimer always told us, "Buy and hold."

We gave our little primer in statistics because we're going to use it to explain some mysteries that you probably didn't even realize were mysteries. We return to Cousin Hermann, who just knows that the universe is out get him.† Of course, he can't think of all the ways that "they" can get him, and to cause trouble Dave gives Hermann one more reason not to sleep at night. "Imagine," Dave says, "you're sitting in your living room, writing quietly, watched by the invisible people who live in your walls, when all of a sudden the air in the room flies right into the kitchen, suffocating you." Scary, isn't it? And technically, not impossible.

"Think about it," Dave says, holding his breath and running outside, leaving Hermann alone in the living room.

Air is made of molecules of oxygen, nitrogen, carbon dioxide, and a few other things, and as these molecules fly around, they almost never hit one another. This isn't a big deal, but it means that as molecules zig and zag around a room, they really couldn't care less what all of the other molecules are doing. Whether a particular molecule is found in the living room or the kitchen is pretty much an even-odds proposition.

If you were the embodiment of the random universe—a "cosmic randomizer"—and it was your job to decide where a particular molecule should be at a particular time, you could do so by flipping a coin. Heads, the molecule is in the living room; tails, it's in the kitchen. At least in principle, couldn't the cosmic randomizer create a vacuum in the living room?

A spontaneous vacuum *could* manifest itself in your living room, but we can say with some confidence that the universe never *will* do such a thing. Here's why it's safe to let your guard down. Imagine there are "only" a million molecules total in the two rooms. In reality there are almost unimaginably more molecules than this, but since we've already run the numbers for a million, we'll stick with that. On average, half of the molecules will be in the kitchen and half in the living room (assuming

---

*This is being written in early 2009, during which the overall trend of the market looks decidedly *non*random.

†What the universe wants with him, though . . . your guess is as good as ours.

both rooms are about the same size). Most of the time, neither room has more than 50.1% of the molecules nor less than 49.9%. These numbers aren't new to you; you saw them before, when Dave and Uncle Louie were flipping coins in the rec room.

Let's face it: it's not a very sizable effect. Keep in mind, though, that these numbers represent the molecules at one instant in time. Just because Hermann is safe at this second doesn't mean that he won't be dead in the next. He shouldn't worry. He won't find himself clutching his throat and gasping for air, even if he lived in the living room his entire life.* If we enlarge the problem to include more realistic numbers of molecules in a room, the density of air in his room won't ever vary by more than one part in a trillion.

When you deal with enormous numbers, such as the molecules in a room, the dictates of randomness are so strict that we might as well say that there is a law. The air will move from a region of high pressure to a region of lower pressure until everything is in equilibrium, for example. But at the end of the day, nothing is deterministic. It's just by far the most likely outcome of the many that could happen.

You don't need to believe us. To put it in terms that you might be more familiar with, let us introduce our nephew, Brian. Brian is a very important young man. He's not simply a teenager living in his parents' basement. He's a *dungeon master* and definitely knows how to work a set of dice. He knows that a standard six-sided die has an equal chance of landing on any given face. That is, you have the same chance of rolling a 6 as you do of rolling a 5, a 4, and so on, all the way down to 1. But what if he were to throw more dice? The chances of the *sum* of the dice having the same distribution will change.

To put this in terms we're sure you're more familiar with, Jeff agrees to play a half-orc barbarian in Brian's latest campaign. To create his character, he picks up a set of three ordinary six-sided dice.† Let's say he rolls them all at once. What's the most likely scenario? Jeff could get a total of 3, but only if all of the dice come up as 1. On the other hand,

---

*Which is ridiculous. Hermann's lived in the attic since he was fifteen.
†If you understand the significance of three six-sided dice, then it's time to roll your charisma attribute again.

it's far more likely that the total of the dice comes up to 10. There are lots of ways for this to happen: 4-3-3, 6-2-2, 6-3-1, and so on. Rolling all 1's is very orderly, since there's only one way for that to happen. Getting a total of 10 or 11, on the other hand, can be done in so many ways that it represents a profound state of *dis*order.

Or to put it another way, there are a lot more ways for a coffee mug to be broken than ways for it to be whole, and many more ways for pool balls to be scattered across the table than arranged in a triangle. Since there are so many more ways for a system—air molecules, an expensive vase, the universe itself—to be disordered than ordered, the natural state of affairs is for things to tend to greater and greater states of randomness. This principle is known as the second law of thermodynamics. It sounds like a guarantee: a system *will* increase in disorder as time goes on.

Physical principles emerge organically from this basic principle. Heat will flow from your piping hot cocoa into your much cooler mouth. That may sound obvious, but the implications are (pardon the pun) truly chilling. For example, the sun and other stars are constantly exhausting their energy, and their heat is flowing into the universe as a whole. Meanwhile, the background temperature is only three degrees Celsius above absolute zero. That means that everything in the universe—planets, stars, galaxies, even Earth—is a red-hot ember, comparatively speaking, constantly spewing heat into space. Since the greatest possible state of disorder is for matter and heat to be spread throughout space, our universe will eventually freeze to death.

We wouldn't tell Hermann, though. He's still terrified to leave the kitchen.* He doesn't need one more thing to dwell on.

 ## How does carbon dating work?

Somehow the molecules in the air randomly bounce around, choosing either to be in Hermann's living room or his kitchen. We just casually alluded to the idea of a cosmic randomizer as though it were the most

---

*Even with Aunt Mavis talking about her bunions and coughing into the bread pudding.

natural thing in the world. But that's just crazy talk. Did you ever wonder how things become random in the first place?

The Cosmic Randomizer

There are two sorts of "random" processes. In one, the system is really deterministic, but you simply don't have enough information, or can't calculate fast enough, to figure out what's going to happen. Take a coin flip, for example. If you wanted to call it in the air and knew the exact position, orientation, weighting, and spin of the coin, as well as the wind direction and speed, then in principle you could run those data through a computer to figure out how the coin will fall. We could repeat the experiment under nearly the same conditions again and again and get the same outcome. If we made a finely tuned robotic coin-flipping machine, we could make heads come up every time.

In practice, there is so much uncertainty with how a coin is held, how the air is blowing, how hard you flip the coin, and where you apply the pressure, that there is absolutely no practical way that you could do these calculations. That's why the coin flip is the random number generator par excellence. Likewise, even though the sequence of cards or the landing of a roulette ball seem pretty random, at the end of the day it's really more that we don't know enough about the initial conditions. Like a giddy dungeon master figuring out damage rolls from a zombie horde, we just can't do calculations fast enough.

But coin flips are not atoms, and something very different happens when we talk about randomness on the subatomic scale. At that extreme, the universe is really, truly random. It's not simply that we don't have enough information. If we ran the movie of the universe forward from identical conditions, quantum mechanics guarantees that we won't get the same results. In the double-slit experiment, the electron really, truly has no idea which slit it will pick before we measure it.

Quantum randomness shows up in all sorts of microscopic phenomena. Most fundamental among these concerns is the radioactive decay of particles, which is how Jeff began discussing this topic with Cousin Hermann. Hermann is as deeply concerned about radioactivity as he is about fluoridated water containing mind-controlling substances. But radioactivity can be used for either good or evil.

Radioactivity occurs because not all atoms are stable. It's a general trend in nature that systems want to devolve into the lowest energy possible. Sometimes, if you leave an atom sitting around for too long, it breaks down into something smaller. Of course, if it loses mass in the process, it gives off its excess mass via a noxious emission called radiation. If that radiation is energetic enough, it can do some pretty nasty damage.

Let's revisit an example we saw in chapter 2 when we talked about tunneling. Given enough time, an isotope of uranium called uranium-238 will decay into a helium nucleus and a thorium nucleus.* On average, this isotope has to wait about 4.5 billion years for half of its atoms to decay (roughly the current age of our sun). This is known as the half-life for radioactive decay. If you start with a block of pure uranium-238, leave it alone for 4.5 billion years, and then see what's left, you'll find

---

*The isotope of uranium in bombs and reactors is called uranium-235, but uranium-238 is pretty unpleasant stuff, too.

that half the atoms in the brick are still uranium and the other half have turned into something else. That "something else" is mostly lead, since thorium itself isn't stable and decays into protactinium in less than a month on average. Protactinium has a half-life of only a few minutes, after which it turns into lead. Lead, in stark comparison, is stable, which means that its half-life is longer than the estimated age of the universe.

The transition from one element to a decayed one is not gradual. On the contrary: when an individual atom decays, it happens in an instant, lasting essentially no time at all. What's more, the uranium doesn't know how long it's been waiting around to decay. There's no time stamp on it. But imagine that the cosmic randomizer looks at one atom of uranium-238 and rolls a 100-quadrillion-sided die once a second. Number after number comes up but nothing happens. Then eventually, completely without warning, a 1 is rolled. That 1 represents a critical failure in stability, and *poof!* we get a decay. If the cosmic randomizer does this for every atom for 4.5 billion years, half of them will still be uranium and half will have decayed, though there's no predicting which will be which.

We can exploit this idea to learn all sorts of things. For instance, carbon-14 is created by cosmic rays in the upper atmosphere, which slowly circulate into our air. All living creatures that process carbon, both plants and animals, take in the carbon-14 (along with the much more common carbon-12). The ratio of carbon-14 to carbon-12 in our bodies (or stems) is fixed at the same ratio as in the atmosphere overall.

That is, until we die.

After death, the carbon-14 starts to decay into nitrogen with a half-life of about fifty-seven hundred years. Anything that used to be alive, or anything made from something that used to be alive, can be sampled. Since the carbon-14 decays, but the carbon-12 is quite stable, by measuring the ratio today and comparing it to the ratio of the two in the atmosphere, we can measure how long something's been dead. Though carbon dating has been used as a very effective tool in archaeology and paleontology, it is fundamentally based on quantum physics.*

---

*And a morbid way to bum out your friends with science.

We asserted, without proof, that the atoms don't know ahead of time when they are supposed to decay and that their decay—along with all other quantum phenomena—is fundamentally random. This uncertainty is deeply unsettling, and the world would be a far more comforting place if there were some way to rejigger the system so we could make the uncertainty go away. Can we?

 ## Does God play dice with the universe?

Einstein didn't like the idea that nature was truly random. You've probably heard his dismissive "God doesn't play dice with the universe." This was a fair bit of scientific conservatism by Einstein and hearkened back to the bad old days of leeches, the aether, and a profound fear of witches. He believed that if we knew the universe in enough detail, we should be able to predict *exactly* how it will evolve.

Einstein helped found quantum mechanics, and it was supposed to have blown away the deterministic edifice, but he was never fully on board. Still, Einstein *was* Einstein, and if he had a problem with your theories, you'd better deal with it posthaste. For a long while, Niels Bohr* was able to answer objection after objection, but by 1935 Einstein, along with Boris Podolsky and Nathan Rosen, his colleagues from Princeton's Institute for Advanced Study, thought he had finally come up with a quantum mechanical paradox that couldn't be resolved. Einstein aimed to prove that God really doesn't play dice with the universe.

This isn't just a philosophical debate. Either radioactive decay or measurements of particle positions are really random, or they just appear that way. The trick, at the time, was coming up with a way of proving the case one way or another. We'll describe the EPR paradox (so named because of Einstein and his gang) in a moment, but it'll be much easier if we make things concrete. Discussing this over Aunt Mavis's ambrosia salad, Brian, Hermann, and Dave (each bringing a bit of nerdy knowledge to the party) decided to make an "entanglement machine."

---

*Founder of the Copenhagen interpretation of quantum mechanics.

Our machine will measure a property common to all fundamental particles called "spin." A spinning charged particle generates a small magnetic field. Because magnets interact with one another, we can measure the direction of an electron's spin by running it past a bar magnet.

Spin is much stranger than you might suppose. If you take an electron with a spin in a particular direction, and rotate it once, it doesn't have the same spin you started with. It's only by rotating it twice all the way around that you return to the original configuration. This is just another way that the quantum world wants to mess with your mind.

The Blombergs can turn the bar magnet into a spin detector, which we can rotate any way we want. If we hold it vertically, we can measure whether the spin is up or down. If we hold it horizontally, we can measure whether the spin is left or right. Spin is particularly useful for our purposes because it's possible to make real systems where two particles are created and the spins *exactly* cancel each other. If one particle has a spin up, then the other must be spin down. If one is left, then the other is right. This is what we mean by "quantum entanglement." It's basically a fancy way of saying that by knowing the quantum properties of one particle, we can say something meaningful about the quantum properties of the other.

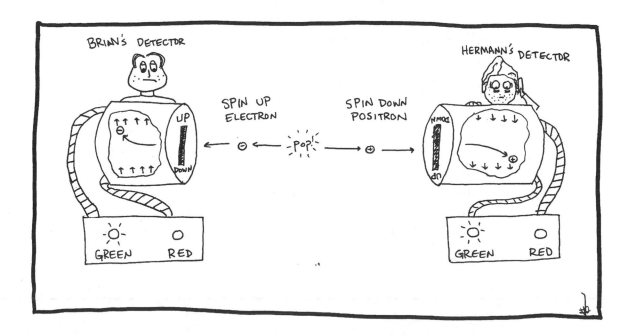

We'll use this property of spin in our entanglement machine. In the middle, we have a chamber where every now and again we make an electron and its evil twin antiparticle, a positron,* set up so the spins of the two are always opposite each other. The electron then travels down a long pipe to the left, where Brian has attached a little detector, while the positron flies to the right, toward Hermann, with a similar detector setup. The detector consists of a magnet, which can be used to measure the direction of the electron's or the positron's spin. To stop us from referring to up, down, left, and right so often, they're rigged so that if the electron is spin up, a green light goes on. If it's spin down, a red light goes on. Hermann's side is set up exactly in the opposite way.

Both of their detectors start off oriented up-down (as shown on page 81), and they shoot out lots of electron-positron pairs. In experiment after experiment, when Brian sees green (the electron with spin up), Hermann sees green (the positron with spin down). Likewise, when Brian sees red, Hermann sees red.

This may not look like such a big deal. We can easily imagine a similar case in which we replace our entanglement machine with something that makes pairs of white and black marbles. If Brian gets the white one, he'd know without looking that Hermann's marble is black, without having to call his uncle to find out for sure. But the Copenhagen interpretation of quantum mechanics is different. In the quantum world, in the instant before Brian measured the spin of his electron, it was both up *and* down simultaneously, and it wasn't until he measured it that it "decided" to be up *or* down. This is where Einstein *finally* gets to make his big objection to the whole thing. By his reasoning, there are really only two possibilities:

1. At the moment it is launched from the central chamber, there's no way in Heaven or on Earth that we can know the spin of Brian's electron or Hermann's positron—even the universe doesn't know.

   But somehow (here's Einstein's big beef) the two particles are able to decide, *at the exact same time*, what they want to be. Let's say that Brian makes his measurement a nanosecond before Hermann.

---

*We'll talk a lot more about antimatter in the next chapter. For now, you can think of "positron" as just a convenient label.

Somehow, in that nanosecond, Brian's electron needs to send a message to Hermann's positron telling it what spin to have. But the electron and the positron are very far from each other, and the two will have to communicate this information almost instantaneously, even if the signal would have to travel faster than light.

This is the EPR paradox. If spin (and any other measurement in quantum mechanics, including the well-being of the cat) is really random, then somehow a signal needs to be sent faster than light. If you got nothing else from chapter 1, your gut should be telling you that this is impossible.

2. Einstein's alternative is that the electron (and the positron) "knew" all along which spin they were going to choose. The only people who weren't in on the secret were the Blombergs, doing the experiment.

Einstein and his crew said that there must be more to reality than the numbers we measure directly. He called this idea "hidden variables," and if it sounds familiar, it should. We saw in chapter 2 that the Copenhagen interpretation of quantum mechanics continued to make people uncomfortable well into the middle of the twentieth century, and Einstein's idea of hidden variables formed the core of Bohm's "causal interpretation" of quantum mechanics. In essence, Einstein is saying that the universe knows the answer; physicists just haven't figured out yet how to get that answer.

The second alternative certainly seems more intuitively correct, and it was Einstein's weapon of choice in the great debate. On the other hand, our intuition has let us down before. We need a way of distinguishing experimentally between the two. Einstein's objections to quantum mechanics remained important but untestable conjectures for about thirty years. In some ways, this is a good thing. It means that for most calculations, either Einstein's hidden variables or the random interpretation could be correct, since both produced the same results.

However, a universe ruled by hidden variables will behave very differently from a random one, and in 1964 John Bell, then at Stanford, came up with a criterion to determine whether the universe is fundamentally random. While "Bell's inequality" is a bit on the mathematical side, we can give you the essence of the test by building a "local reality machine," for which you can find the design specifications in the box at the end of this chapter. If you want the Cliffs Notes version, the idea is that if Brian and Hermann orient their detectors randomly and run their electron/positron generator many, many times, Einstein's hidden variables picture suggests that they'll see the same light turn on in their detectors more than half the time. The Copenhagen picture of quantum mechanics, on the other hand, predicts *exactly* half.

Even though we had the math, for almost twenty years actually doing the experiment was technologically out of reach. It wasn't until 1982 that we had our answer when Alain Aspect and his collaborators built a contraption strongly resembling the reality machine and put the EPR paradox to the test. They found that the lights matched exactly half the time. In other words, the Copenhagen interpretation of quantum mechanics won out. The electrons did not behave as if they'd been preprogrammed, as Einstein had hoped.

This has some bizarre implications. It means that by measuring the spin of an entangled electron, the corresponding positron is *forced* to be in the opposite spin state *faster than the speed of light*! Crazy, you say? Einstein even referred to this as "spooky action at a distance." Don't fret, though. We don't need to throw the (relativistic) baby out with the bathwater. In fact, we only need, very gently, to adjust our rule about nothing traveling faster than light. Since you can't use quantum entanglement to send messages across the universe, we just have to add that nothing *carrying information* can travel faster than light.

God, it seems, really does play dice with the universe. But to us, the *biggest* uncertainty is whether we'll make it out of our family reunion alive.

Mermin's Local Reality Machine

While the actual derivation of what's come to be known as Bell's inequality is on the mathematical side, the main point can be made without any advanced math and provides the basis for the Cornell physicist David Mermin's hypothetical "local reality machine." This reality machine requires a simple adjustment to Brian and Hermann's electron generator and will allow us to figure out, once and for all, whether the EPR paradox disproves quantum mechanics. You're just going to have to do a bit of counting, and the only question you're going to have to ask is, "Does something happen more than half the time?"

Assume for the moment that Einstein was right that inside each electron is a miniature program. No matter how Brian and Hermann orient their detectors, the program will tell the detectors which lights to turn on, and the program has to account for all eventualities. For instance, a particular electron might turn on the green light if the detector is oriented vertically, and the red light if the detector is horizontal. The positron must have the same program.

We're going to need to tweak our generator so Brian and Hermann can turn their detectors into one of *three* possible positions when measuring the electron or positron.

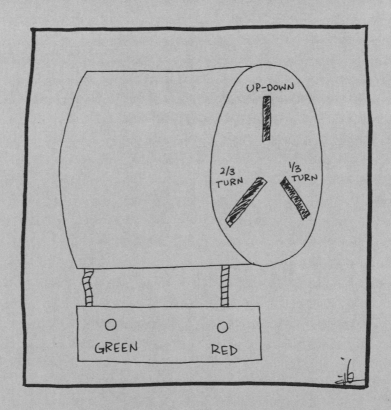

Brian and Hermann can set their detector in one of three positions: (A) up-down, (B) turned a third of the way around, or (C) turned two thirds of the way around.

We pick these particular options because quantum mechanics makes a very specific prediction. If we run the reality machine again and again, and each time Brian and Hermann choose the position at random, quantum mechanics says that their lights will match *exactly* half the time, on average.

We know that "exactly half" must strike you as coming from out of the blue, and we apologize. For the most part, we want you to have an intuition about things, but in this case, the "half" comes out of a fairly complicated quantum mechanical calculation, and we ask you to take our word for it.

What do Einstein's hidden variables predict? Here you don't have to take our word for it at all. There are only eight possible programs that the electron can be set to have:

	(A) Up-Down	(B) 1/3	(C) 2/3
1	G	G	G
2	G	G	R
3	G	R	G
4	G	R	R
5	R	G	G
6	R	G	R
7	R	R	G
8	R	R	R

Remember how these programs work. If Einstein is right, then no matter how Brian or Hermann decide to turn their detectors, the electron needs to know ahead of time which light is going to turn on. There are essentially eight different kinds of electrons that can be produced.

Only two of the variables from each program can be measured at any one time, since we have only two detectors. This tells us what Brian's or Hermann's light will say for whichever position they pick, respectively. So, for example, if Brian turns his dial to the up-down position (A), and he sees a green light, he doesn't know whether the electron was preprogrammed with GGG, GGR, GRG, or GRR. But the universe does!

The interesting result from the reality machine comes when we look at what happens when an electron and a positron with a particular program are shot out. If Brian and Hermann randomly pick which direction to turn their dial, how often will they get the same colored light?

There are two easy programs: GGG and RRR (cases 1 and 8). No matter what happens in those cases, Brian and Hermann will get the same readings on their detectors. "Always" is definitely more frequent than "half the time."

A more interesting case is GGR. There are nine different ways in which Brian and Hermann can turn their dials: A-A, A-B, A-C, B-A, B-B, B-C, C-A, C-B, and C-C. We have to list them all we're afraid because the point is kind of subtle. Of these nine scenarios, in five of them (A-A, A-B, B-A, B-B, and C-C) Brian and Hermann will see the same lights. Five out of nine is about 56%, which is more than half.

The other six possible programs—GRG, GRR, etc.—are all exactly the same as GGR, since two of the slots are set the same and one is different. In those cases, Brian and Hermann will still get the same signals 56% of the time.

No matter how the electron is programmed, in Einstein's model, Brian and Hermann will get the same signals *more than* half of the time. On the other hand, if quantum mechanics is correct, they'll get the same signals *exactly* half the time.

This ends in tears for Einstein.

The Standard Model

"Why didn't the Large Hadron Collider destroy Earth?"*

Neutrinos interact weekly.

---

*If, by publication date, the LHC actually *did* destroy Earth, then you have our most sincere apologies. We will happily offer a full refund for this book.

On March 21, 2008, Walter Wagner and Luis Sancho filed a lawsuit in U.S. federal court with one simple goal: save the world. The threat, they claimed, was that the Large Hadron Collider (or LHC, for short) was set to come online in the next few months, and once it did, it would produce microscopic black holes that might coalesce and ultimately eat our Earth from the inside out.

They weren't alone. As physicists, we're asked all the time whether the LHC will destroy Earth, or maybe even the entire universe. All of this hounding has made us feel somewhat paranoid about playing some role in dooming humanity. Online petitions abound, calling to shut down the CERN (translated from the French, the European Organization for Nuclear Research), the LHC parent organization. Some were written with compelling, rational pleas for research and precaution. The majority of available online petitions, though, seemed to be written by furious children in a series of cell phone text messages:

> On September 10th 2008 CERN (European Organization for Neclear Research) will set off its first beam in a machine called The Large Hadron Collider. Now; this is for a scientific reason, I do not know. If succesfull im shure many of scientist's

questions may be answered. But! If the LHC is succesfull. We may never know the answers to the questions. [*sic*]

Even Nostradamus, in an only marginally less coherent way, wanted in on the action, offering *this* less than friendly shout-out from the past:

IX-044	IX-044
Migrés, migrés de Geneue trestous,	Migrate, migrate out of Geneva everyone,
Saturne d'or en fer se changera:	The sky will change from gold to steel:
Le contre Raypoz exterminera tous,	The Antichrist will exterminate all,
Auant l'aduent le ciel signes fera.	Before the event the sky will show signs.

Of course, he failed to predict that Ang Lee's film *Hulk* would suck, and *everyone* saw that coming. On the surface of it, the LHC *does* look like a doomsday device: it's a giant underground ring seventeen miles in circumference—so large, in fact, that it crosses the French-Swiss border four times. You can think of the LHC as a light-speed monster truck rally in which particles are accelerated up to 99.999999% of the speed of light* and then smashed into one another. As we saw in chapter 1, energy and mass are interchangeable, so at these tremendous speeds a bunch of high-mass particles are created. The LHC is the biggest advance in particle collisions in recent history, and the fear is that one of the things created by these collisions might spell doom for humanity.

Except that it hasn't.

First, while "particle accelerator" sounds scary, this is not new technology. If you've ever used an old-style television, you've seen a simple particle accelerator at work. The old sets used cathode ray tubes to

---

*This is not hyperbole, as if we're saying "a gajillion." We've seen that nothing can travel faster than light, but that doesn't stop us from building accelerators to get *very* close.

accelerate electrons, and by adjusting the position of the beam, magical moving pictures were created on your screen. The mechanism inside the LHC is somewhat different, but just as with television, particle accelerators can be illuminating and terrifying.*

Early Particle Colliders

So which is it? Is the LHC another important step toward our ultimate understanding of nature or, like Icarus, are we flying too close to the Sun? Will we be punished for our arrogant pursuit of knowledge?

Rest assured, nobody is in danger of losing an eye. How do we know? Settle in for a while, because before we figure out why the LHC doesn't pose any danger, we have to figure out why we're building it in the first place.

---

*For example, much has been learned about the human threshold for pain, based solely on the fact that *Small Wonder* ran for four straight seasons.

What do we need a multibillion-dollar accelerator for, anyway?

In high school physics, everything seemed like a hodgepodge of arbitrary rules: do *this* calculation if you have a pulley; do this other thing if you have an inclined plane; do this third thing if there is acceleration; and on and on, down the line. Honestly, trying to keep track of the rules of motion and friction is probably what turns people off physics[*] in the first place.

That's a shame, because physics is not as monstrous as people make it out to be. The goal of physics is to have as few rules as possible. However, this isn't to say that if we knew those simple laws, then doing physics *calculations* would be simple. Assume for a moment that you had a friend[†] who had never seen a chessboard before. You could describe the rules of chess to this friend in just a few minutes. He could observe a chess game, note that everything happens in accord with the rules, but he still may not be able to play very well.

We started this chapter with a heavy vibe, speculating on the possibility of destroying the world. We'd like to lighten things up by thinking of physics as a game—or a set of games—along the lines of tennis or badminton. These games seem very similar to each other on the surface. In each case you have two or more players hitting a ball (or shuttlecock) back and forth over a net, with the basic goal of trying to make the other person or team miss.

The object is to figure out the rules of the game, figure out which players can play the game, and maybe say a thing or two about the ball. Ideally, we will eventually show that all of the seemingly disparate games are, in fact, one super-awesome metagame, kind of like the decathlon. Physicists have done a pretty good job describing these physical laws by breaking them down into two parts:

---

[*]And, for similar reasons, dating.
[†]If you try really hard, we know you can do it.

1. The players: There are a bunch of fundamental particles.
2. The games: There are four forces, each of which has its own fairly similar set of rules. Not all of the particles play all of the games.

Collectively, this set of particles and rules is known as the Standard Model. The Standard Model serves not only as a description of the *stuff* that made the universe, but also as a convenient title with an exploitable pun.

THE STANDARD MODEL IS
OF GREAT INTEREST TO
PHYSICISTS.

Let's start with the basics: All matter is fundamentally made of atoms.* This idea has been around since at least 1789, when the chemist Antoine Lavoisier posited that you couldn't slice up stuff with infinite

---

*All of the matter you can see and feel, at least. When we talk about dark matter, all bets are off.

smallness; eventually you'll reach the smallest possible particle.* These "indivisible" particles have come to be known as atoms, but it's only been in the past century or so that we've had an idea of how small and compact atoms really are.

In 1909, Ernest Rutherford conducted an experiment in which he fired a beam of what were known as "alpha particles"† toward a thin piece of gold foil. Most of the alpha particles went clear through the foil without being deflected at all. Every now and again, though, the alpha particles bounced backward. In Rutherford's own words, "It was almost as incredible as if you fired a fifteen-inch shell at a piece of tissue paper and it came back and hit you." This, of course, was the basis of the very first hyperbolic textbook cover intended to make physics "come alive."

What Rutherford found was a tiny speck smack dab in the center of the atom. This blob is what we call the nucleus, and when we say it's little, we mean it. Given the gargantuan scales we're going to be using in our discussion of cosmology and the submicroscopic scales we're using here, it might be easier if we use "scientific notation": the nucleus is about 10^{-15} times the volume of the atom. That's a factor of 0.000000000000001. To put things in perspective, that's roughly the same as comparing the volume of a house to that of the entire Earth. Since 99.95% of the atom's mass resides in the nucleus, it's fair to say that the atom is overwhelmingly filled with empty space.

Small as it is, even the nucleus is not fundamental. If you bust your way into a nucleus, you'll find even *smaller* particles, known as "hadrons," though you may know them by their individual names: protons and neutrons. The protons, in fact, are the buggers that get smacked into one another in Geneva, giving the Large Hadron Collider its name.

---

*Oh, we know. In the fifth century B.C.E. Democritus and Leucippus came up with the idea of "atoms" as being the smallest, indivisible things in the universe. However, any similarity between their atoms and the ones we now know is purely coincidental.

†Particle-naming sounded much more science-fictiony back then. Since we had no idea what matter was made of, we had names such as "alpha particles," "beta particles," and "gamma rays." These have been replaced, respectively, with "helium nuclei," "electrons," and "high-energy photons." The old days seemed much cooler. This explains the steam-punk movement.

The two hadrons would be nearly indistinguishable aside from two important facts: the neutron is .01% more massive than the proton, and the proton has an electric charge of +1, while the neutron is electrically neutral—hence its name. We'll worry about the implications of electric charge in a bit, but sufficed to say, if you've ever worn a wool sweater on a dry winter's day, you're probably already aware of it.

We've accounted for about 99.95% of the mass of an atom with our hadrons, but still haven't said anything about the tiny remainder, the component that apparently fills up the vast majority of the volume. It's a little something called the electron, which we first described in chapter 2. This time around, we want to talk about electrons as "fundamental" particles. No matter how you pry or peel them, they just don't break down into anything smaller.

To put electrons in perspective, they're as common as protons and neutrons, but for a 150-pound person, only about half an ounce of

Ask Dr. Science: Can We Build a Shrink Ray and Make Miniature Atoms?

The vast majority of the volume in atoms is made up of empty space. Sure, there's a nucleus and an electron. Still, as we saw in chapter 2, an electron isn't like a ball bearing or the flesh of a peach (with the nucleus playing the part of the pit). It's a big probability wave. Couldn't we just build a ray or a contraption to make the electron clouds smaller? Things wouldn't be any lighter, of course, but our shrink ray would make packing for long trips as easy as pie.

The problem that we run into is one of uncertainty. As we saw in chapter 2, when you try to confine an electron to a small volume to make super-tiny atoms, Heisenberg's Uncertainty Principle says that the energy of those electrons goes way up. The energy would get so high that the electrons would escape the electromagnetic pull of the nucleus.

At the end of the day, the size of the atom comes out of a pretty straight-forward combination of physical constants: the charge of an electron, Planck's constant (the number that tells us how strong quantum mechanics is), the mass of the electron, and the speed of light. If we could rejigger the fundamental constants of physics, *then* we could make miniature atoms. Until that day, it's probably easier to buy a bigger suitcase.

their weight comes from electrons. If you were to suck all the electrons out of your body and then suck out both of your eyes, they would weigh about the same. Like protons, electrons also possess an electrical charge, though unlike protons, theirs is -1. In normal atoms, protons and electrons are found in equal numbers, which equates to electrical neutrality.

Neutrality isn't just limited to Switzerland and atoms. However matter was created in the universe, there are exactly as many positive as negative charges in the universe, and so the entire universe is, and always has been, electrically neutral. There isn't a single experiment that has been done, terrestrially or otherwise, that doesn't conserve charge. This leads into our first ground rule for all of the fundamental forces:

Electrical charge can be neither created nor destroyed.

As you might expect, the action in our universal game consists of more than just shuffling around protons and electrons from place to place, conserving charge all the while. As an example, let's look at a neutron. A neutron is kind of like a patient in a doctor's office: after an average of ten minutes or so waiting, the neutron flies to pieces. The exception, though, is that instead of screaming at a medical assistant, a neutron literally blasts apart, and a bunch of other particles come flying out.

The biggest thing to free itself is a proton. This may come as a surprise, since we told you that electrical charge has to be conserved, but this isn't an issue if there is some other particle with a *negative* charge to cancel out the positive charge of the proton. Something like an electron. *Exactly* like an electron.

A few other things come out of a neutron decay as well, but we want to make two caveats now: (1) Despite appearances, a neutron isn't made of a proton, an electron, and some other stuff; it turns into them. (2) On a related note, the protons and neutrons *are* made of *something*, we just haven't said what.

We'll get into the other fundamental particles in a bit, but before long you're likely to get overwhelmed by the "particle zoo." We don't want to make you memorize a big catalog of fundamental particles, for the simple reason that there are at least eighteen of them, not including weird variants of the same particle, which don't really differ from one another in any fundamental way. As a service to you, we've made a handy appendix at the end of the chapter that contains everything you might want to know about the particle zoo. Really, there's no need to thank us.

You now know about as much about what matter is really made of as anybody did a century ago, but we're going to have to dig a bit deeper to figure out what's going on at the deepest levels. This is why we want to smash the ever-living snot out of the particles at the LHC. Our hope is that protons are like piñatas or members of the AV club: if we hit them hard enough, something interesting will come out.*

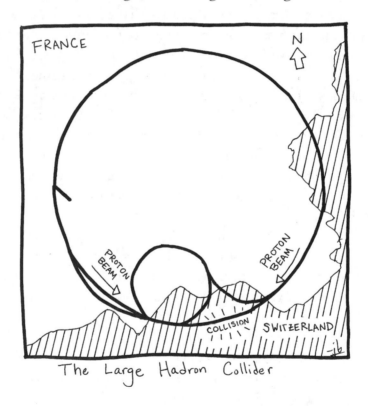

The Large Hadron Collider

The circular ring is the proton racetrack of the accelerator and has two beams of protons flying at each other at nearly the speed of light. As we saw in chapter 1, it takes a heck of a lot of energy to get particles

---

*It's all in good fun, folks. One of us (Goldberg) was on the "Mathletics" team through most of high school, so who are we, really, to judge?

moving this quickly. Chugging through the numbers, using the same amount of energy it takes to accelerate two protons to speeds high enough to thoroughly destroy them, we could *make* fourteen thousand of them from scratch using $E = mc^2$. Once two protons smash into each other, lots of things can happen, but all of them are governed by the second of our ground rules:

Energy can neither be created nor destroyed.

It *can*, however, be converted from motion into mass, and that is what we aim to do in our particle colliders.

 ## How do we discover subatomic particles?

Smashing our energetic protons into one another creates particles much more massive than those we started with. But if the particles created in accelerators are so massive, why do we need the accelerators at all? Shouldn't gigantic particles be easy to spot?

Yes and no. Sure, if there were massive particles floating around in space, we could pick them out without a lot of trouble. The problem is that everything in the universe wants to drop to lower energy if at all possible. Put a bowling ball on a table—a position giving it fairly high energy—and give it a little nudge. It will fall off the table and onto your foot, to a lower energy. Since energy and mass are equivalent, this also means that a massive particle will decay—if at all possible—into a *less* massive one, and in very short time, as we saw when we talked about radioactivity in chapter 3.

Most massive particles last for only a millionth of a second or less before they decay into something lighter, and so presumably 13.7 billion years or so after the beginning of time all of those massive particles will have done their decaying once and for all. You might assume that everything will settle down to the common protons and neutrons that we know so well, but you know what happens when you assume, don't you?

High-energy charged particles are constantly shooting around the universe. From the Sun, from other parts of our own Galaxy, from supernovas, any place that has a high-energy source can shoot protons out at high speeds. These charged particles, called cosmic rays, zip around until they hit something. If not for the magnetic field that surrounds our planet, that "something" could be your cells, sterilizing or killing you. This is why you should listen to the advice your momma gave you: don't spend extended periods in space. Often enough, cosmic rays hit our atmosphere and collide with oxygen or nitrogen, becoming a more massive particle in the process. Like a mouthful of unbrushed teeth, the stratosphere and everything higher is alive with gunk: particles such as muons, kaons, and pions.

These particles are born and die in the blink of an eye,[*] so the only way to create and measure them in any useful way is inside an accelerator. If we smash particles together at high enough energy, then invoke $E = mc^2$—voilà! We get out massive particles. By producing them in accelerators, we can more easily predict when they will occur, which makes them easier to study.

But pions and muons are not the only massive particles to suffer from degenerate tendencies. As we've already mentioned, even the neutron is susceptible to decay, a trait it does not share with the proton.[†] If you give it about ten minutes, a neutron will decay into a proton, an electron (so that the total charge is conserved), and another particle that we didn't tell you about previously, an antineutrino.

Don't freak out just yet, because we're going to explain both the "anti" and the "neutrino" part. Let's start with the neutrino. The name comes from the fact that neutrinos are electrically neutral, and can't be seen directly. How did we even know that they were there if they were essentially invisible? Clever guesswork.

In 1930, Wolfgang Pauli put forth a novel interpretation of neutron decay experiments. It had been noticed that when a neutron decayed, the proton and the electron sometimes flew off in a similar direction. Like so many things in life, Pauli's interpretation of neutron decay can be made clearer by invoking superheroes.

---

[*]Technically, about a hundred million pions could lead a rich and full life in the amount of time it would take you to blink your eye.
[†]Maybe. We'll pick this question up again in chapter 9.

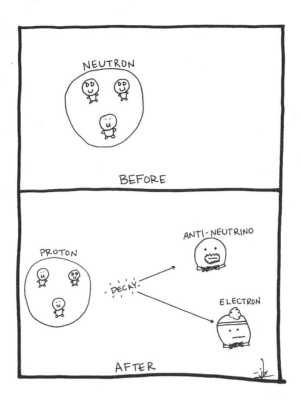

Imagine that Sue Storm (aka the Invisible Woman) and her husband, Mr. Fantastic,* are skating on a frozen pond. The two push off each other, and Mr. Fantastic goes flying off in one direction, and Sue, invisible as ever, goes flying off in the other direction. The Thing, watching the scene from the bank, sees only Mr. Fantastic, sailing backward, seemingly without cause. But he figures it out pretty quickly. He *knows* that there must be someone else—someone invisible—flying off in the other direction.

Pauli (playing the part of the Thing) realized that there must be an unseen ghost particle that is electrically neutral, the antineutrino.

Neutrinos (and thus antineutrinos) are very light, and for a long while they were thought to be completely massless. However, in 1998, the Super-Kamiokande experiment in Japan showed that neutrinos

---

*The Fantastic Four actually got their powers from cosmic rays, making this example doubly relevant.

have some mass. While this is an impressive achievement, it should also be noted that at the moment, physicists haven't actually measured what the masses are. We'll return to this question in chapter 9, but for now all we can say for certain is that the mass is many times smaller than that of an electron.

As far as the "anti" part, try not to be spooked by the name. Anti just means opposite: an antiparticle has the exact opposite quantum numbers of its partner particle. Antimatter is one of those substances that has gotten a bad rap, since everybody knows that if a blob of anti-matter comes into contact with ordinary matter, then the two blow up and release all of their mass into energy. Antiparticles themselves are harmless. If we suddenly took all of the particles in our universe and replaced them with antiparticles (including the ones making up you), you'd never know the difference.

Why are there so many different rules for different particles?

Now that we've set a few ground rules common to all the fundamental forces, it's time to talk about the games, starting with the most obvious.

Gravity

For the record, *of course* people were aware of the existence of gravity before Sir Isaac Newton "discovered" it in 1687. People were able to make catapults work, for instance. They were aware that if they shot an arrow up, it would eventually pierce armor, hopefully on the other side of the field. Without gravity, the Halifax gibbot, a predecessor of the guillotine, would have simply sat, its blades occasionally flying out of the top of the machine.

But with a simple set of equations, Newton was able to *predict* with great accuracy the falling of an apple, the orbit of the Moon, and the paths of the planets. His law was simple; it explained an enormous range of phenomena. His law showed that all objects in the universe attract one another gravitationally, and the farther they are from one another, the weaker the attraction.

Newton didn't get the whole story, though. It wasn't until Albert Einstein developed his theory of general relativity in 1916 that we were really able to understand the gravitational force. It's only when we start talking about black holes (chapter 5), the entire universe (chapter 6), or the Big Bang (chapter 7) that we really need to worry about where Newton went wrong. But for our purposes here, he was correct *enough.*

We said before that each of the forces is a lot like some sort of two-player ball sport. If we had to pick, gravity would be kind of like badminton. It's played over a large field (the whole universe, in fact), and is very low-impact. You can imagine getting hit with a shuttlecock, and, compared to the kind of things you can get struck by in other sports, it won't leave a lasting impression.

It's a great starter game, because not only is it safe for all ages, but everybody's eligible to play. All particles, massive or otherwise, create gravitational fields and are attracted to one another.

Electromagnetism

Unlike gravity, which is always attractive, electromagnetism can be either attractive or repulsive. You already know that particles can have one of three types of electric charge: positive, neutral, or negative. Electrons, when placed side by side, always repel one another. A positively and a negatively charge particle pair, like a proton and an electron, always attract. If either particle is neutral, they do nothing at all.

Electrons are repulsive.

While two electrons attract each other gravitationally, they also repel each other electrically. We have an unhealthy competitive streak as much as the next guys, so we'll ask the question on everyone's mind: between gravity and electrical forces, which one dominates?

Electricity wins, and not even by a three-point shot at the end of the last quarter. It's a total rout. The electrical repulsion between two electrons is more than 10^{40} times larger than the gravitational attraction, and *that*, more than anything else, is why we can afford to ignore gravity when we talk about atomic scales and smaller.

You'll note that we're referring to this force as electromagnetism, but so far we've only talked about the "electro" part. From a casual perspective, electricity and magnetism seem very different, but at a fundamental level, the differences are just a matter of perspective. Just as stationary charges make electric fields, moving electrical charges create magnetic fields—that's how an electromagnet works and how we understood spinning charged particles in chapter 3. Likewise, changing magnetic fields can create electrical fields, which in turn create electrical currents.

The astonishing thing is that electromagnetism explains virtually every physical phenomenon in your everyday life. It's electrical repulsion that prevents your rear from falling through your chair. Electrical attraction holds molecules together and is the basis for all of chemistry. And yes, static electricity is what holds a balloon to a wall.

And what of magnetism? Apart from bar magnets and MRIs, we don't see too much of this in our everyday lives. But it is extremely important in particle accelerators. When a charged particle (such as a proton) travels through a magnetic field, it moves in a circular orbit. The stronger the magnetic field, the faster the circular orbit. By putting a bunch of magnets in the LHC ring, a beam of protons will be able to be trapped at speeds close to the speed of light.

Electromagnetism is like tennis. It's a lot more fast-paced than other sports, and the little fuzzy green balls pack a heck of a punch. Neutral particles can't play because they aren't "seen" by the photon and because they left their racket at their mom's house.

All *charged* particles can play the game of electromagnetism.

The Strong Force

It was necessary to introduce electromagnetism because there were observed phenomena, such as the existence of atoms and molecules, that couldn't be explained by gravity. Together, though, gravity and electromagnetism still don't explain everything.

Consider helium. It's made of two neutrons and two protons. As far as electromagnetism is concerned, the neutrons basically sit this one out, but the protons *really* don't want to be in close quarters with

one another. Within the nucleus of every single helium atom, the electrical repulsion between the protons is about fifty pounds! Why doesn't its own electromagnetic repulsion blow helium apart?

There must be another force that acts on both protons and neutrons and forces them to stick together. This is known as the strong force, and it only comes to play on very, very small scales—about 10^{-15} meters. We throw around these numbers a lot, but to put it in perspective, comparing the width of the atomic nucleus to your height is about the same as comparing your height to the distance from here to Alpha Centauri.

As it turns out, the rabbit hole goes even deeper. In the 1960s, the Deep Inelastic Scattering Experiment at the Stanford Linear Accelerator fired high-energy electrons into atoms. The resulting ricochet showed that there was something else inside the protons and neutrons—protons and neutrons were not fundamental particles, but were composed of something smaller. These smaller somethings are known as quarks.

Along with electrons and neutrinos, quarks are the final players in our metaphysics game. There are six different types of quarks (and you can look at their adorable mugs in the appendix to this chapter), but only the lightest two concern us for now: the up quark (with an electric charge of $+2/3$) and the down quark (with an electric charge of $-1/3$). Protons have two ups and a down,[*] and neutrons have two downs and an up.[†] What holds the whole thing together is the strong force. The strong force is so strong, in fact, that quarks are *never* seen outside a proton or a neutron.

The strong force is a lot like table tennis. Confined to close quarters, it is an intense head-to-head competition. Only quarks (and protons and neutrons, which are made of quarks) can play the game of strong force.

The Weak Force

When we introduced the strong force, we claimed that we had to do so because there were some unexplained phenomena that the other two forces (gravity and electromagnetism) couldn't adequately explain. We've already mentioned one: neutron decay. We said that left to their own devices, a neutron will decay into a proton, an electron, and an antineutrino. Try explaining *that* using one of the forces we've already talked about!

We have to invent (okay, hypothesize) an additional force, and since we've already used all of the good names, we've settled on "the weak force." The neutrinos are pretty much the signature particle of the weak force, since (being neutral) they sure as heck can't play the electromagnetism game, and only quarks play the strong game. Except for this small difference of charge, it turns out that neutrinos and electrons are a lot alike, and the weak force, among other things, allows neutrinos to turn into electrons and vice versa. Trillions of neutrinos pass through you every second. They are produced in the Sun by the quadrillions, and yet giant detectors count only a few of them every day. This rarity of interaction is a sure sign that the weak force is well named. And since neutrinos interact *only* via the weak force, you don't see too many of them directly.

---

[*]Up + up + down = $2/3 + 2/3 - 1/3 = 1$. Add up the charges of the quarks and you get the charge of the proton. Pretty cool, huh?
[†]We'll let you do the math on this one.

The weak force is a lot like tossing around a medicine ball. It is very close-range, low-impact, and on a typical time scale, incredibly boring. Actually, we already have a hint as to *why* it's so boring. The medicine ball is heavy, and even the stereotypical olde-tyme mustachioed musclemen couldn't throw it very far.

The Weak Interaction,
as Medicine Ball

Quarks, neutrinos, and electrons all get to play in the weak nuclear game. Sure, everyone gets to participate, but as we've already said, it's a pretty slow game, and not a lot happens.

 ## Where do the forces *really* come from?

We started this discussion by saying that the fundamental forces were like games, but we're still missing one key component to make our game any fun: the ball. Think about it. Without a ball, tennis is just a game of swinging a racket spastically. The same is true in particle physics. As we currently understand it, if we put two electrons on a table, they'll just sit there. They interact only by communication through the electromagnetic (or weak, or gravity) field. Without the field, they can't see each other.

Where does that field come from? The two particles must somehow alert each other to their presence. This can be achieved by "sending" a third particle between the two of them. This messenger, or *mediator*, is the particle that actually carries the force. The two electrons send a particle back and forth with the message "I'm here. Get away!"*

The fuzzy green mediator particle in our electromagnetic tennis game is called the photon, and we've already spent a fair amount of time talking about it in chapter 2. We already know they're massless and travel at the speed of light. As a consequence of the vacuum energy permeating the universe, we are constantly awash in photons that pop in and out of existence.

As we saw, depending on the circumstances, light can be described as a particle or as a wave. More generally, a wave is just a kind of field—something you can measure everywhere in time and space. If you moved an antenna around your house, you would detect different radio signals—sometimes weaker, sometimes stronger. This is the electromagnetic field. A photon is just a little packet of this electromagnetic field speeding through space at the speed of light. The same will be true for all of the fundamental forces. There is a strong field, a weak field, and a gravity field, and each of these has a corresponding particle.

For the strong nuclear force, the mediators are known as gluons. Like photons, gluons are massless, and travel at the speed of light, but unlike photons, gluons suffer from separation anxiety. While the photon is the carrier of the electromagnetic force, the photon itself is electrically neutral. It doesn't *feel* the electromagnetic force.

Particles that experience the strong force have a different kind of charge that goes by the name color. Red, blue, and green are the strong force analogs of negative and positive in the world of electromagnetism, and dictate the interactions that quarks can have in the strong field. If you

---

*Or, sometimes, "Do you find me attractive?" The sad, eternal answer is, of course, "No."

got your crayons out to draw strong interactions, forget it. It's just one of those wacky naming conventions meant to confuse the layfolk.

There is an important distinction between the electromagnetic regime and the strong regime, though. Like with electromagnetism, the "players" (quarks) have charge, but unlike with electromagnetism, so does the ball. Not only do gluons *carry* the strong force; they also *feel* it, in stark contrast to photons. Gluons attract one another and get all tangled up into structures called glueballs. This means that gluons can't travel very far before being tripped up; this is one of the main reasons why the strong force is confined to the nucleus. That goes double for quarks, which give recluses such as J. D. Salinger and Thomas Pynchon a run for their money. They are *never* seen outside the nucleus.

Our theory of gravity, called general relativity, doesn't require any mediator particles. We'll talk more about general relativity in chapters 5, 6 and 7, but the fact that relativity seems so different is a mystery that will presumably be resolved when a "Theory of Everything" (or a convincing one, at any rate) gets developed.

If all of the forces are "really" the same, then shouldn't they all have a mediator particle? The idea is that gravity is carried with a particle called a graviton, but not only has this never been detected, we're nowhere near building an experiment sensitive enough to find one. However, we do know that if gravitons are real, then, like the photons, they must be massless. That's why they are able to send gravity signals over enormous distances.

The weak force seems very different from the other fundamental forces, because it's the only one with *three* mediator particles. Unlike the cool names given to the other mediators, they are simply called the W and Z bosons.* Why is the weak force so weak, and why does it require subatomic distances to have any effect? We've already seen the answer. They're massive, like medicine balls, and it's very hard for them to travel over large distances. That may seem like no big deal to you, but according to the simplest theories, the weak force, like electromagnetism, and actually all forces, should have a massless mediator particle. Why are these particles different?

---

*There are two different *kinds* of W's. That's how we get up to three.

In the physics world, it's not okay to be different. Physicists *really* love symmetry. They pass notes to symmetry in study hall and bring it flowers after school. What physicists generally mean by symmetry is that you can change something about the system without the underlying physics getting altered.

Imagine that you went out for a day of minigolf with your niece and nephew, and subscribing to traditional gender roles, you give a blue golf ball to your nephew and a red one to your niece. When you start the round, it doesn't matter who has blue and who has red, since they work exactly the same.

Now imagine that halfway through a hole, you distracted the kids with some delicious soft serve and then switched the red ball for the blue. If you tell the kids that you've switched them, there's no problem. They can easily pick up where they left off, with your nephew now whacking around the red ball and your niece the blue. Of course, you couldn't switch only one ball, leaving two red balls on the course. In that case, they wouldn't know which ball to hit, and you'd have ruined a nice day out.

Let's make this a bit more scientific than putt-putt for a moment. Symmetry is important because fundamentally, any two electrons— or any two fundamental particles of the same type—are exactly the same. At the microscopic level, there is no such thing as, say, *this* electron or *that* electron. We simply note that there are two of them.

Almost. Electrons have another property, called *spin*, as we saw when we talked about the EPR paradox in the previous chapter. The spin of an electron can be up or down. What's the difference? In many ways, there's no difference at all. An up-spinning electron has the same mass and charge as a down-spinning electron. On the other hand, if we pass a spin-down electron through a magnetic field, it gets deflected in a different direction than a spin-up electron does. Moreover, a magnetic field can be used to change a spin-down electron to a spin-up electron, and vice versa. This is where the symmetries come into play. Physicists note that two particles are pretty much the same, except for some relatively small difference. We think of them as two versions of the *same* particle.

Of course, sometimes the analogy gets particularly strained. For example, in minigolf you can always switch the red ball for the blue without a problem. They play exactly the same way. But what if we switched the red ball for a bowling ball? This switch would be a bad symmetry from the perspective of playing golf, since one ball can fit into the hole and the other can't. However, if you weren't playing golf, but you wanted to see if a floor was level, presumably a bowling ball or a golf ball would do the trick equally well.

On top of all that, electrons have another property, called their phase, which can't be measured. The only thing that can ever be measured is the *difference* in the phase between two electrons.* Two electrons with different phases are in some ways the exact same particle, and in some ways they are different.

Electrons can be a real pain in the neck.

In the 1940s, Richard Feynman of Caltech figured out a whole new way of looking at the whole shebang. He asked what would happen if there was a field that could change the phase of an electron (or any other charged particle) to some other phase. Crunching through the math, he found that the field is *exactly* the one for electromagnetism. This weird assumption—that electrons with one phase can be turned into electrons with another—formed the basis for predicting everything about light. Had he done the calculation forty years earlier, he could have predicted photons before Einstein proved they existed.

We totally acknowledge that this approach, which is called Quantum Electrodynamics (QED), seems completely made up. We have absolutely no idea why our universe chooses to make its physical laws in such a way that this symmetry argument works. But it *does* work.

Now is when, more than ever, physicists like to think back to their old friend under the bleachers: symmetry. If this approach works for one fundamental force, might it work for the others? On the face of it, neutrinos and electrons don't look much alike. The electron is negatively charged, while the neutrino is electrically neutral. From the perspective of electromagnetism, the two are very different indeed. While both

---

*Think of "phase" like the vertical hold on an old-fashioned TV set. You can still make out the picture, even if it's rotated upward a bit.

THE STANDARD MODEL   113

particles are extremely light, neutrinos are so puny that for a long time physicists assumed that they were completely massless.

However, there's clearly something that connects electrons and neutrinos. If a neutrino comes out of a reaction, you can bet your bottom dollar that there was an electron involved there somewhere. So perhaps there's symmetry between the two, albeit a very weak one. The assumption is that there is a weak field—actually three of them—that can turn an electron into a neutrino or vice versa, or an up quark into a down quark, or allow neutrinos to bounce off one another. Little "blobs" of the field could be detected as the W and Z particles.

We could go through the same sort of much more complicated argument and figure out the properties of gluons, the carrier of the strong force, or of the hypothetical graviton, the carrier of gravity. We won't, however. We (and the investigators at the LHC) are interested in solving a mystery in the weak force. As with electromagnetism, the weak force equations that come out when we do our symmetry calculation work nearly perfectly.

Nearly.

In chapter 1 we saw another form of symmetry. We didn't call it that then, but we noted that all of the physics in the universe made sense whether you were traveling at constant speed or standing still. We also saw that the apparent speed of particles changed depending on whether you were moving or standing still. There was one exception: massless particles always move at the speed of light.

There is clearly something special about massless particles, and the upshot is that using our symmetry arguments, we would expect every one of the mediator particles to be massless. Photons and gluons are. Even though we've never detected gravitons, the fact that gravity travels at the speed of light means that gravitons have to be massless.

The W and Z particles, on the other hand, have lots of mass.* They're about a hundred times as massive as a proton. In the true spirit of mathematics, we're going to have to do some *serious* fudging of the equations to fix this.

---

*When they sit around the house, they sit *around* the house.

Why can't I lose weight (or mass)—all of it?

According to our best guess, the symmetry arguments we described before really do describe the fundamental equations in the universe. Particles really can turn into one another. If that guess is correct, we could predict each of the fundamental forces, the existence of electrons and neutrinos, the different kinds of quarks, and so on.

Except that we can't. Like a sumo wrestler on a pogo stick, the problem here is mass. It's not just W and Z particles that should be massless. If we were starting from scratch, making the simplest possible model of the universe, we'd have assumed that the quarks, electrons, and neutrinos should be massless as well. They're not.

Most popular books on physics talk about concepts such as "spontaneous symmetry breaking" and other technical terms with an eye toward describing mass in real particles. These are actually code to describe the mathematics used to (ahem) correct the equations, to make them predict what we actually see.

Now, we don't want to go too far. There's nothing dishonest about this. In fact, it's science of the best sort. You come up with a theory, the universe doesn't happen to abide by your prediction, so you invent a new tool to correct the mathematics. Quarks were invented as a mathematical tool initially, but then it turned out that they just happened to actually exist.

It would be silly to describe the math required to get around the problems we've encountered so far. It would *not* be silly for us to get to the bottom line. In the 1960s, Peter Higgs of the University of Edinburgh proposed that there might be yet another field in the universe beyond the mediator fields we've already talked about. In true creative fashion, it is known as the Higgs field. There is one huge difference between the Higgs field and the others previously mentioned, though: the Higgs field doesn't carry a force.

The Higgs field is everywhere. In fact, you're soaking in it. But if the Higgs field is surrounding us, why don't we notice it? What does the Higgs field do? In the simplest interpretation, you might think of it as being like molasses. Put a quark in a big bucket of Higgs field and give a little shove. What happens? The quark, interacting with the Higgs field, is harder to push than you might think. Physically, the more difficult it is to move something, the more massive it is. In that sense, the Higgs field "gives" the particles masses.

We don't want to take this analogy too far. If the Higgs field really were molasseslike, then a particle, once in motion, would start to slow down. This clearly doesn't happen. Still, the basic picture is that just as the electromagnetic field creates an interaction that moves charged particles, the Higgs field creates an interaction that gives a particle mass.

This seems like we're just making things up, doesn't it?

But this isn't just a case of a twitchy scientist grasping at straws. We mentioned the idea that the various forces in the universe might be just different aspects of a single force. Historically, for instance, electricity and magnetism were thought to be very different, until 1865, when James Clerk Maxwell showed them to be just different aspects of a single electromagnetic interaction.

Since then, physicists have been trying to show that the remaining four forces are really just three, or two, or ideally one. What does that mean? After all, the fundamental forces sure *look* different. That's true today, but as it turns out, it all depends on whether the universe is hot enough for you.

In 1961, Sheldon Glashow, Steven Weinberg, and Abdus Salam showed that electromagnetism and the weak force were one and the same. This seems like a tall order. The differences between the weak force and electromagnetism are pretty staggering. Electromagnetism has a massless mediator particle, and the weak interactions all take place through W and Z particles—particles that are very, very heavy. As a result of this, electromagnetic interactions can take place over a great distance, while weak interactions are all very short-range.

By now, you get it: these forces are different. It's *weird*. So how can two seemingly separate things be united? Glashow, Weinberg, and Salam looked at the forces as they may have looked in the early universe, at high temperature and energy. They found that a complete theory of electroweak interactions would have four mediator particles, all interacting with more or less the same strength.

As the universe cooled, however, the Higgs field (which had been around all along) started to get tired. And as it (metaphorically) settled into retirement, it started getting involved in its community. Three of the electroweak particles (the W's and the Z) started interacting with the Higgs, giving them mass, while the photon continued on being massless.

It seems like a nice story, except for one little thing. We need a reason to believe that the two very distinct forces could be combined. The electroweak theory isn't infinitely malleable. We can't just make up any old story and hope to make it stick. One of the most solid predictions of electroweak theory is the ratio of the masses of the W and Z particles. The Z particles were predicted to be about 13% heavier than the W's—a prediction that has since been experimentally verified to ridiculously high accuracy.

The hitch in the plan is that for all of this to make sense, the Higgs field has to exist. Otherwise, the electromagnetic and weak fields would still be joined. The other option is that the theory is way off base, and we have to start from scratch. However, for our collective sanity, we'll assume for the moment that the Higgs field is real. In that case then just like all the other fields, a little blob of Higgs field should be able to be observed as a real particle. The only problem is that the Higgs particle is electrically neutral (meaning that it's hard to detect under normal circumstances), and extremely massive, which means that it's hard to make in a collider and, once made, it decays really quickly.

We don't know exactly how massive it is, but if it were really light, then we would have detected the Higgs particle already, and if it were too massive, then the W and Z particles wouldn't have the mass ratios they do. These two constraints put some hard limits on the Higgs mass from about 120 to 200 times the mass of a proton, and the name of the game—besides just finding the Higgs—is to figure out what its mass is. Even before the LHC, physicists at the Fermilab Tevatron collider showed in early 2009 that the mass *can't* be between 170 and 180 times the proton mass.

How are we actually going to get one of these bad boys out of the colliders? Up to this point, we've talked about colliding proton beams, but it's a lot more interesting than smashing one proton into another. When the particles get accelerated, they gain quite a bit of energy.

But when two protons meet, it's not the protons themselves that hit; it's the squishy stuff inside that collides.

The quarks and the gluons inside each gain a lot of energy from the trip around the collider, and it's the gluon-gluon collisions that unleash a huge amount of energy to construct giant particles like the Higgs.

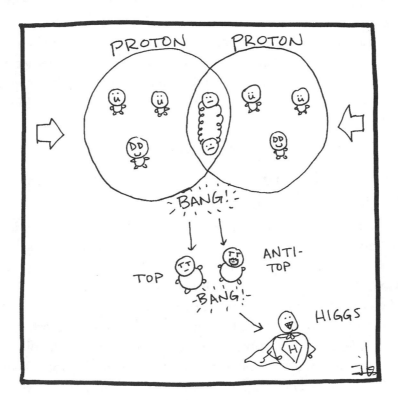

We're making a lot of this up, or at least making a very sketchy guess based on what we know. We do know that these particles have never been detected in any particle accelerators, but the LHC represents the highest-energy experiments we've ever performed. This means that while we've already covered the lower end of the mass range in previous accelerators, we'll finally be able to probe the highest Higgs mass predicted by theory. And we're confident that if we smash two quarks together at high enough energy, a Higgs will pop out of the reaction.

If it exists.

How could little ol' LHC possibly destroy the great big world?

We have some idea of why the LHC was built, but we also know that curiosity killed the cat.* It's great if we discover the Higgs. It would certainly prove that we're clever, but we'd hate to be too clever for our own good.

For example, if we can get one massive particle—the Higgs—out of smashing two quarks together, might we be able to get another, more different and dangerous one? Certainly lots of stuff can be created in high-energy collisions. The fear is that when two particles collide, they might create some pretty scary stuff: a black hole or some exotic matter called "strangelets." Might one of these destroy the world?

Mega-Deadly Scenario 1: Earth Is Swallowed from within by a Black Hole

We'll talk a lot about black holes in chapter 5, but for now you need know only one important fact: if you drop your keys into a black hole, let 'em go, because, buddy, they're gone. There's a point of no return known as the "event horizon," and the more stuff that falls in, the bigger the event horizon, and hence the black hole, gets.

So what would happen if two protons smashed into each other in the LHC and were somehow turned into a black hole? The black hole would be at most about fourteen thousand times the mass of a proton—pretty puny by normal standards. What's more, the event horizon would be many times smaller than even the scale of an atomic nucleus. You'd have to have *very* good aim to get even a particle to fall into it.

You might feel that this will allow you to let your guard down, but don't be foolish! Remember, our microscopic black hole is an unstoppable killing machine. If it encounters other particles, it'll swallow them and grow faster and faster. The fear is that the microscopic black hole would form, start to grow, fall to the center of Earth, where it would continue to grow, and eventually swallow the whole planet.

Pretty scary, huh?

---

*As you may or may not have seen in chapter 2.

The LHC and its parent organization, CERN, were so concerned about the PR problems surrounding the LHC that they set up two assessment groups, one in 2003, and another in 2008, to figure out if the world was actually going to be destroyed. Their conclusion was "there is no basis for any conceivable threat from the LHC." Well, *of course* you'd expect *them* to say that! But if we think about it a bit, we're going to come to the same conclusion.

The first piece of comfort comes from the fact that in a sense, the entire LHC run has already been performed on Earth—about a hundred thousand times, and we are still here to tell the tale. Cosmic rays move at even higher energies than the ones produced in the LHC. And they hit the atmosphere constantly. Any dangers that could come from high-energy protons hitting one another have been repeated again and again.

Earth still exists; ergo, the LHC also won't destroy Earth.

Let's forget about the fact that Earth hasn't yet been destroyed, and consider *why* it hasn't. Consider, first of all, that despite the enormous energies in the LHC, we can only produce particles up to a particular mass. As we already said, the upper limit is about fourteen thousand times the mass of a proton. In practice this limit is smaller, because it's really the quarks and the gluons that are colliding, not the entire proton. In reality, only particles about a thousand times the mass of a proton will be created.

On the other hand, if we understand anything about how the universe works, the minimum mass of a black hole is about twenty billionths of a kilogram, or what's known as the Planck mass. The Planck mass may seem small, but it's about a quadrillion times larger than the most massive particles popping out of the LHC.

Where does this limit come from? It comes from uncertainty. In chapter 2 we saw that we can't say with absolute certainty *where* a particle is, and the less massive the particle, the greater the uncertainty. On the other hand, when we talk about black holes, what we mean is that all of their mass is confined within the event horizon. The upshot is that if a black hole is too small, then it won't all "fit" within the event horizon. The crossover point comes at the Planck mass.

Everything we know suggests that black holes smaller than the Planck mass can't form. But what if we're wrong, and they form *anyway*?

In chapter 5 we'll see that all black holes eventually go kaput. The smaller the black hole, the quicker the evaporation. It's almost pointless to talk about how quickly an LHC black hole would evaporate (provided one could even form in the first place). To put it in perspective, from the time when the black hole formed to when it disappeared, it would be able to travel only a microscopic fraction of the size of an atomic nucleus. In other words, it wouldn't have time to absorb anything.

What's more, one way or another, we're pretty darn certain that the black hole *will* evaporate. If we've learned one thing from particle physics, it's this: if you can create a particle in a collision, then the particle can decay as well.

Mega-Deadly Scenario 2: Strangelets Form and Grow into a World-Strangling Crystal

In most of our discussion, we've focused on the operating mode in which the LHC will collide individual protons into one another. On the other hand, there's another mode in which they will collide individual nuclei of heavy atoms—lead, mostly—into one another, and this ion mode has evoked a whole additional set of fears.

You might think that we've already covered all of the bad things that can happen. Lots of the cosmic rays that hit our atmosphere all the time are, in fact, made of heavy ions. How is that any different from what's going on in the LHC? The difference is that the heavy ions in our atmosphere are hitting light things, such as oxygen, nitrogen, and hydrogen, and so on Earth we've never really seen what'll happen when we smash two bits of lead into each other.

We have, however, seen what will happen on the Moon. After all, the Moon doesn't have an atmosphere, and cosmic rays bombard the Moon all the time. The Moon, it's fair to say, hasn't been destroyed, so we probably should feel safe as well.

You're not convinced, and we hear you. "Safe? Safe from *what?*"

To answer that, we first need to point out that there are more types of quarks out there than the up and the down quarks that we talked about already. There are a total of six different types, of which the up and the down ones just happen to be the lightest. The next lightest is known as the "strange" quark, and like the down, has a charge of $-1/3$.

We've already commented that by and large, heavy particles will decay into lighter ones if given half a chance. Strange quarks are no different. However, it's possible that "hypernuclei" made with one or two strange quarks will actually be *lighter* than the normal nuclei. Don't believe us? In an ordinary proton, it turns out that only 2% of the mass comes from up and down quarks. The rest comes from energy—the energy of motion of the quarks and the interaction energy between the quarks and the gluons.

It may be that hypernuclei could form "strangelets" (made up of roughly equal numbers of stranges, ups, and downs) in the LHC. This is all kind of speculative, since strange quarks aren't around long enough to do any real experiments with. We don't actually know *what* would happen if you started injecting strange quarks into ordinary matter. As a result, lots of different theories abound.

A few of these theories are downright apocalyptic. The fear is that once you have one strangelet, it will bind to ordinary matter, and the ordinary matter will catalyze into the lower-energy strangelet matter. This would continue indefinitely, basically destroying the planet and everything on it. Incidentally, this is almost exactly the doomsday scenario envisioned in *Superman Returns*, albeit with strangelets replaced by kryptonite.*

This would be terrifying—except for the fact that strangelets don't appear to exist. The Relativistic Heavy-Ion Collider at Brookhaven National Labs collides heavy ions, as you might expect. They have seen no evidence for strangelets. Likewise, strangelets don't seem to form from collisions of cosmic rays.

So rest easy. While physics may yet produce devices destined to destroy Earth, this giant hole in the ground isn't it.

---

*If this sort of destruction appeals to you, you may also be interested in Kurt Vonnegut's *Cat's Cradle* and a handful of Valium.

If we discover the Higgs, can physicists just call it a day?

We've been on pretty solid ground so far about what we expect to find in the LHC. The vast majority of physicists would be very, very surprised if the Higgs particle *isn't* found. What we know for sure is that the LHC is not the end of the world, and the Standard Model is not the end of the story. Here's a little taste of things to come.

String Theory

Whether you're a diehard physics nut or, conversely, you have a social life, you've probably heard of something called "string theory." String theory was developed as a way of explaining several mysteries[*] we've been skirting thus far. Gravity is very, very different from the other three fundamental forces in the universe.

The weak, strong, and electromagnetic forces *require* a mediator particle—and in each case, we've discovered those particles experimentally. Our theory of gravity, general relativity, not only doesn't require a graviton, but so far, no graviton has been found. What's more is that it seems strange that the theory behind the particles of matter (quarks, electrons, and so on) should be so different from the theory of the force carriers (photons, gluons, and the like). What we'd love is a theory of unification—ideally, a Theory of Everything, or "TOE," as the cool kids say.

Though not yet fully developed, string theory is a leading contender for a TOE, and its central player, as you might imagine, is called a string. Think of a string as a rubber band, a very tiny one, perhaps as small as 10^{-35} meter around. And what are these strings, really? Simply put, they're everything.

---

[*]Mrs. Ethel Kranzton, an eighty-one-year-old nanny from Belding, Michigan, independently developed string theory as a way of explaining the unsightly knots in her cross-stitching. It is largely accepted by quiltworkers, but rejected by scientists as "convoluted knit-picking, and mathematics that are needlelessly difficult."

You need to realize that all of the particles we talked about in this chapter—quarks, electrons, photons, and so forth—are treated in the Standard Model as being infinitesimally small. They are literally points. The Standard Model doesn't really explain why one particle has one mass, and one charge, and whatever other properties it has, and another particle has the properties *it* has.

String theory says that the only reason why these particles look like points is that we're not looking closely enough. In reality, "point particles" are really tiny loops that are constantly vibrating. If this idea sounds familiar, it should. It's exactly what we saw in quantum mechanics when we saw all sorts of things—photons, electrons, vacuum fields—oscillating back and forth.

The more vigorously a string is vibrating, the more massive it is; remember, $E = mc^2$ can be run in reverse. The other properties of the oscillation determine everything else about the particle. To account for all the properties of the particles we actually see, strings can't just wiggle about in the three dimensions we're normally aware of. That doesn't mean the strings can't exist, it just means we need more dimensions.

Don't get us wrong. We can't move into these higher dimensions. For one thing, we'd be a bit cramped. Many, perhaps all, of the additional spatial dimensions are very small, far beyond anything we'd be able to detect in the LHC. Even if we were able to travel along these hidden dimensions, they'd behave a bit like a Pacman universe,[*] and we'd return to where we started in no time at all.

There's just no way to make string theory consistent with the laws of physics in our universe using only three dimensions. In theory after theory, the number of potential dimensions grew and grew, until Edward Witten of the Institute for Advanced Study at Princeton proposed the current leading contender in 1995. His version, known as M theory, posits that we're living in a whopping ten-dimensional universe.

---

[*]For all you kids out there, Pac-Man was an awesome video game from 1980 in which you play a yellow circle with a mouth who eats smaller, whitish circles. If you disappeared through a tunnel on the left side of the screen, you'd reappear on the right. Also, there were ghosts.

In many ways, string theory looks very promising. It provides a framework in which all four of the fundamental forces can be unified into a single theory. It describes the forces and the particles as just different sides of the same underlying physics. It might even provide insight into the nature of space and the beginning of the universe, as we'll see in chapters 6 and 7, respectively.

On the other hand, there are a few problems. First, it is *very* difficult to test string theory. Because the scales of strings are so small, we have little hope of being able to probe string theory using the LHC or any experiment that we're going to build anytime in the foreseeable future. The other problem is that string theory doesn't address all the unanswered questions in particle physics.

Loop Quantum Gravity

There is another big hole in the Standard Model, one that string theory doesn't even pretend to answer. How do we reconcile the two great theories of the twentieth century, quantum mechanics and general relativity, our theory of gravity? Both theories are "correct" in that they tell us what happens in the realm of the very small, and in the realm of very strong gravity, respectively. But they can't both be right. What happens in environments such as black holes or near the beginning of time, when both theories are expected to put in an appearance?

Think about it. As we saw in chapter 2, uncertainty seems to dominate nearly every aspect of physics—the vacuum energy of photons, the motion of electrons, the paths of photons. Quantum mechanics is hardwired into the three nongravity forces. The similarities among them constitute the reason why the electromagnetic and the weak forces can be thought of as a single electroweak force. It's also the reason physicists have proposed a number of competing Grand Unified Theories (GUTs) to combine the electroweak with the strong force. Gravity is different. As weird as it is, general relativity doesn't have any of the randomness that shows up in the other three forces. What we'd really love is a theory of quantum gravity.

One of the most exciting and potentially fruitful approaches is known as Loop Quantum Gravity, or LQG for short. One of the strangest features of LQG is that space itself is quantized. That is, if you look on small enough scales, space would no longer seem smooth but would appear pixilated. We never notice this under normal circumstances, because the scales we're talking about here are about 10^{-35} meter, a distance known as the Planck length. The Planck length is as much smaller than an atom as an atom is smaller than the distance to the nearest star outside our solar system. Space literally can't become any more compact than this. This is going to have some interesting implications when we discuss the Big Bang in chapter 7.

One of the appeals of LQG is that it doesn't require any more than the three dimensions we're normally used to, plus one for time. It also gives rise naturally to the graviton, making our picture of particle physics far more unified. On the other hand, LQG is not, in and of itself, a Theory of Everything. The other force laws need to be put in by hand, as do quarks and the other fundamental particles of matter.

All of this physics beyond the Standard Model may seem like a transparent ploy to keep us in business long after the last particles have collided at the LHC. True enough. But did you really think violence would solve all our problems? Like it or not, it will take more than a few high-energy explosions to reveal all the secrets of the universe.

APPENDIX
Rogues' Gallery of the Fundamental Particles

Throughout this book we've tried to keep lists of things as brief as possible. The Standard Model of particle physics is astonishing because the list of particles (while lengthy) is so simple. The matter in the universe consists of two fundamental types, leptons and quarks. Each group is further subdivided into three generations, and within each generation there are two particles, one with more negative charge than the other. We break our list down by generation, and you can see that all particles seem to have certain things in common. You should also use this as a handy guide to interpreting the cartoons.

The Leptons

Name	Electron	Muon	Tau
Charge	−1	−1	−1
Mass	0.026% of a proton	11.3% of a proton	190% of a proton
Discovered	In 1897 by J. J. Thomson	In 1936 by Carl Anderson	In 1975 by Martin Perl by colliding electrons and positrons

These particles are the charged leptons. They keep their charges in their hats. Because they are charged, they interact with the electromagnetic force. All leptons also interact with the weak force, and all particles everywhere interact with gravity (so we won't mention it again). The electron is the only one that we normally see. The muon decays in about a millionth of a second, while the tau lepton disappears even more quickly.

Name	Electron Neutrino	Mu Neutrino	Tau Neutrino
Charge	0	0	0
Mass	?	?	?
Discovered	1956 by Clyde Cowan and collaborators	1962 by Leon Lederman and collaborators	2000 at Fermilab, in Batavia, Illinois, by the DONUT collaboration

These guys have no hats and hence no electrical charge. If they look similar to one another, that's no surprise. The various types of neutrinos can turn into one another without warning (merely by exchanging their ties), and seemingly without any form of interaction. This "neutrino oscillation" (confirmed in 2003 at the KamLAND detector near Toyama, Japan) actually means that the neutrinos *must* have mass. How much? It's very hard to say, but the upper limits on the electron neutrino are less than 0.3% the mass of an electron. The limits on the other two, however, are much higher, and the tau could be as high as thirty times that of an electron according to current measurements. On the other hand, it could be much, much lower.

The name of each neutrino comes from the fact that they are most strongly associated with the decays or interaction of the electron for the electron neutrino, the muon for the mu neutrino, and the tau lepton for the tau neutrino.

You may notice in the neutron decay cartoon on page 101 that the antineutrino has a goatee. That's just our homage to the classic *Star Trek* episode "Mirror, Mirror" (season 2, episode 33), in which the evil "anti-Spock" always sported extra facial hair. The same is true for all of our antiparticles.

The Quarks

Name	Up	Charmed	Top
Charge	+2/3	+2/3	+2/3
Mass	~0.4% of a proton	~130% of a proton	~180 times a proton
Discovered	1967 at the Stanford Linear Accelerator (SLAC), deep inelastic scattering experiment	1974 by Ting and Richter, independently	1995 at the Fermilab Tevatron facility

These particles are the positively charged quarks. They look very similar, except that the later generations get plumper and plumper. The top quark is the meatiest particle yet discovered. He's just bursting at the seams. He was also the most recent particle to be discovered.

We'd be remiss if we didn't point out a mystery here. You'll note that the up quark has a mass of only about 0.4% of a proton. This is kind of odd because a proton is made of two ups and a down. You'll note that at most, the quarks add up to about 1% to 2% of the requisite total. Where does the extra mass come from?

The extra mass comes from energy. The quarks (as well as the gluons) are flying around pretty quickly and interacting very strongly, and just as mass can be converted to energy, energy can be converted to mass. If you thought it strange that the Higgs could "create" mass, then think of this as just another case in which $E = mc^2$ can be exploited in reverse.

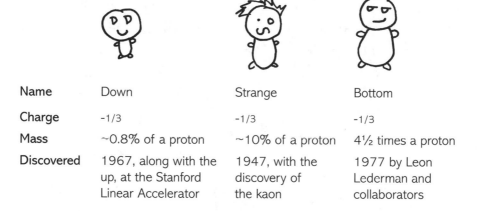

Name	Down	Strange	Bottom
Charge	-1/3	-1/3	-1/3
Mass	~0.8% of a proton	~10% of a proton	4½ times a proton
Discovered	1967, along with the up, at the Stanford Linear Accelerator	1947, with the discovery of the kaon	1977 by Leon Lederman and collaborators

These are the negatively charged quarks. Strangest among them is the strange quark. When particles called kaons were discovered in 1947 they seemed completely nonsensical. They decayed into particles like anti-muons and neutrinos, but were so massive (about half the mass of a proton) that they were inconsistent with any particle yet known.

It wasn't until Murray Gell-Mann proposed the idea of the quark in 1964 that it was understood that kaons were made up of an *anti*strange quark and either an up or a down. Stranges have the distinction of being detected before we knew what they were.

The Mediators

Name	Photon	Gluon	Graviton
Charge	0	0	0
Mass	0	0	0
Discovered	1905 by Albert Einstein	1979 at the German Electron Synchrotron by the TASSO collaboration	Stay tuned.

These particles are the massless mediators—the carrier particles for three of the fundamental forces. It's kind of strange to list the discovery date of the photon, since we "detect" them constantly. However, Einstein's 1905 interpretation of the "photoelectric effect" marked the first time we understood light to be carried by particles. Gluons were detected only in the past thirty years or so.

Gravitons, the carriers for the gravitational field, not only haven't been detected, but according to general relativity, they aren't really needed. However, there's ample reason to suppose that gravity should be like the other fundamental forces, and therefore a mediator should exist.

Name	Z^0	W^+	W^-
Charge	0	+1	−1
Mass	97.5 times a proton	86 times a proton	86 times a proton
Discovered	1983 at CERN by the UA1 collaboration	1983 at CERN by the UA1 collaboration	1983 at CERN by the UA1 collaboration

These roly-poly particles are responsible for carrying the weak force. You will note that except for their hats, they all pretty much look alike. This is no accident. In fact, the W$^+$ and the W$^-$ particles are so closely related that they are antiparticles of each other. One of the great triumphs of twentieth-century theoretical physics was the calculation of the ratio of the Z/W masses, about 1.13. This is a direct prediction of the Higgs model and has since been measured experimentally with extraordinary accuracy.

And our hero:

The Higgs. He's chargeless but certainly not charmless. He's the only particle in the Standard Model that hasn't yet been discovered, so we don't know exactly how massive he is. Our best guess is that he's 120 to 200 times the mass of a proton. Because he interacts strongly with massive particles, he has an on-again, off-again affair with the top quark.

Time Travel

"Can I build a time machine?"

Have you ever wanted to ride a dinosaur? Have tea with the tsar? Make a killing in pork belly futures? Or, if you're the killer robot type, have you ever wanted to prevent the birth of the one person who can stop your robot insurrection? You're going to need a time machine, and they don't come cheap. In our view, you're far better off building your own, and while we won't stand in your way, we imagine your family won't be too thrilled. They'll tell you it's impossible. They may even accuse you of being mad.

But are time machines impossible? And what's so bad about being mad?

There are worse things in the world than being mad, especially if you're a scientist. Regular scientists may be able to ping voltages using a cathode-ray oscilloscope, but mad scientists can *stop time with a freeze ray.* There are heroes to sabotage, and attractive heroes' girlfriends to kidnap. If we had to do it over again, we probably would have majored in mad science over the regular version.

As a standard piece of sci-fi–level mad technology, consider teleportation. As we saw in chapter 2, this comic book standard is already in our grasp. Unfortunately, at the moment, we're only able to move one atom at a time, and even then, it's really just easier to carry the damn thing.

The point is that while the gadgets of comic books and science fiction don't necessarily defy science fact, oftentimes they're just not worth the effort. Perhaps this is one of the reasons why mad scientists have so much trouble. Another reason might have something to do with the fact that most of the devious equipment that gets cooked up violates some very serious laws, and not just the ones enforced by superheroes, or meddling kids and their dog.

Can I build a perpetual motion machine?

Consider the old classic the perpetual motion machine. This staple of mad science is a contraption that never loses any energy, doesn't wear down, and runs forever.* The best of them go one better and continuously *produce* energy, apparently out of thin air.

Journal editors really like to get submissions involving perpetual motion machines because they require minimal effort on their part. "Nope," they say, "conservation of energy† says you can't get something for nothing." They may be paraphrasing the actual conservation law, but they have the feeling right: locally, energy can neither be created nor destroyed, and the energy within a closed system may be transformed (to or from mass, for example), but the total must stay constant.

Maybe the mad scientists with dreams of perpetual motion or energy-making machines are all fools. After all, the types of people to reveal their master scheme to the only person who can foil it are the same types of people to overlook little snags such as energy conservation. But then it could be possible that they have found a loophole in the rules written on the fabric of space and time.

Sometimes it's hard to tell the mad scientists from the regular ones. To drive home the very point we're now trying to make, Richard Feynman, then at Caltech, came up with a pretty clever—but intentionally flawed—perpetual motion machine. Would you like to hear

---

*It was a long-held belief by at least one of the authors that Dick Clark, host of the New Year's Rockin' Eve, was in fact a perpetual motion machine.

†A principle so basic that it's now known as the first law of thermodynamics.

how it works? *Of course* you would. To help demonstrate, we introduce you to a couple of criminal masterminds with evil scientific charisma leaking out their ears: Dr. Dave and his partner in crime, Robo-Jeff.

1. Dr. Dave takes a laser beam and shoots it upward, toward the top of a cliff, where Robo-Jeff is waiting with a collecting dish.

2. After collecting the light, he can turn the light into mass (never mind the particulars of how) using Albert Einstein's great relation $E = mc^2$.

3. Robo-Jeff drops the mass off the cliff. As you know, when you drop something, it gains energy.

4. Voilà! When it gets to the bottom, there's more energy than they started with. They can put some of the energy back into the laser and use the rest for something useful, such as powering a *bigger* laser.

A Perpetual Motion Machine?

The only problem is that it doesn't work, as Feynman knew from the outset.

We haven't figured out a way to violate the first law of thermodynamics, and what we've shown from this whole contraption is that as light escapes from a source of gravity, it *must* lose energy. If a laser beam is shot up a cliff, the energy of the beam at the top must be less than the energy at the bottom. On the other hand, as light falls toward Earth, it must gain energy. This isn't just fancy speculation. In 1959, Robert Pound and George Rebka, then at Harvard, were able to measure the loss of photon energy as photons flew up the side of Harvard's Jefferson Laboratory—a scant seventy-four feet high.

This loss of energy is not easy to measure. In the Pound-Rebka experiment, the photons lost only one quadrillionth of their initial energy. Even if we were to fire a laser up a cliff so tall that it extended into deep space, we'd lose only one billionth of the energy. Unsurprisingly, this is not the sort of thing we notice in everyday life. If gravity were stronger, it would be much more noticeable and much easier to measure.

For an excellent example of tight little bundles of gravity, we turn to white dwarf stars. White dwarfs possess a mass about a million times greater than Earth's even though they are of comparable size, so gravity is about a million times stronger. If you were on a white dwarf, you would be a million times heavier; and if we were of a weaker temperament, we would be making fat jokes.

But there are even more extreme environments than white dwarfs in the universe. Imagine that we stand on the surface of a very, very massive planet where the gravity is very, very strong, and we shoot a laser up into the air. As the photon flies higher and higher into the air, it loses more and more energy.

Now imagine that the planet is really, *really* compact. In that case, the light will lose so much energy that it will reverse direction and return to the surface of the planet. Or will it? If the planet really were so dense that light couldn't escape, it would never move upward in the first place. It's like a little kid trying to walk up a down escalator. Bless his heart, he's trying, but he's inexorably sliding down and down. In fact, such a "planet" wouldn't even have a surface in the first place.

That, too, would collapse under the immense gravity, and the whole planet would collapse to a single point—a singularity.

Making one of these singularities isn't easy to do. To generate such a gravity using our own Earth, we would have to smoosh the mass so small that the resulting object would be about a third of an inch in diameter. Even the Sun, three hundred thousand times more massive, would have to be compressed down to less than two miles in radius to trap light. That's smaller than the borough of Manhattan.

This is the basic idea of a black hole—a system so compact that light itself cannot escape. The point of no return, the "event horizon," is an invisible boundary between the hectic pull of very strong gravity and a one-way ticket to the center of a massive beast. Once anything—a star, a lone sock, a lunch box, a particle—crosses the event horizon, it is dragged into the black hole. Even photons are powerless to escape from its greedy maw. Since light can't escape once past the event horizon, neither can anything else. Remember: the speed of light is the speed limit of the universe.

Black holes seem like an essential tool in the mad scientist arsenal. They can be used for any number of things, from disposing of pesky protagonists to ditching failed biological experiments. But what the really mad scientist wants is somehow to exploit the time-warping properties of gravity near a black hole to build a time machine.

Before we get into the hows and the whys of a black hole, and if you can (or can't) make a time machine out of one, we want to remind you of a few features of photons, the particles that make up light that we discussed in chapter 2.

As you'll recall, if you've seen one photon, you've seen them all. The only real difference is that some are more energetic than others. There are a bunch of properties of light that might seem different at first but are actually all of a kind. In the case of light, the amount of energy a photon has correlates to the color it appears. This interdependence of energy and color goes beyond the range to which our eyes are sensitive.

In chapter 2 we also discussed how light behaves like little wave packets and that higher energies mean a shorter wavelength. The most important point (for the purpose of this discussion) is that since photons are little waves, we can time how long it takes for successive wave fronts to pass a fixed point, a time interval known as the period of the wave. Remember in chapter 1 when we talked about a cesium clock? We're now ready to tell you what we were really talking about. If you take a photon emitted by cesium and measure the time between crests of the wave, then it behaves like a little clock—one of the best clocks in the universe.

For longer wavelengths of light (lower energy) the crests come relatively slowly. A radio wave, for example, beats about a hundred times every millionth of a second, which is an eternity to subatomic particles. For shorter wavelengths, the period is shorter as well. Knowing only these few facts, and armed with our laser-guided thought experiment, we're almost prepared to rediscover one of Einstein's great triumphs: general relativity.

Are black holes real, or are they just made up by bored physicists?

General relativity tells us how gravity *really* works, and correctly describes the gooey innards of things such as black holes. Among other things, we'll see that time and space aren't nearly as absolute as we'd thought and that near black holes things get really, really weird.

Imagine that Dr. Dave and Robo-Jeff were to take their perpetual motion machine to a planet with very strong gravity. They once again fire their laser up the height of a cliff. By the time the laser reaches the top, it's lost some energy and gotten a bit redder. By extension, the period of the photon measured at the top of the cliff will be longer than at the bottom.

This is the photon version of our cesium clock! Now let's put it to use. Say Dr. Dave sent a beam of photons up a cliff with a period of one second (made of really low-energy radio wave photons). If the planet has a strong enough gravity, Robo-Jeff, waiting at the top of the cliff, might see crests every *two* seconds.

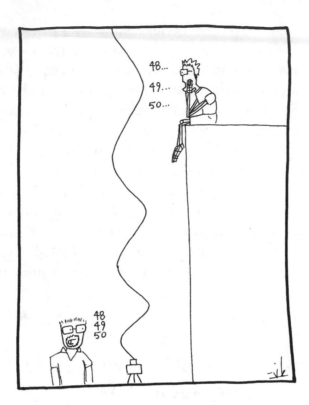

Here's where things get screwy. If we're setting Dr. Dave's watch at the bottom, we notice that after fifty seconds, we'll see fifty crests go by. However, at the top of the cliff, Robo-Jeff would see only *twenty-five* crests pass by over the same period of time.

How can that be?

The only explanation is that time is passing more slowly for Dr. Dave than for Robo-Jeff. Think about it: Dr. Dave's clock appears to Robo-Jeff to run slower by a factor of two, so Dr. Dave is aging half as slowly. Like in our discussion of special relativity, this is *not* an optical illusion. Dr. Dave is aging slower, his digital watch is ticking slower, and he seems to be moving in slow motion when seen by Robo-Jeff.

This time difference is true in general. Clocks run slower nearer massive bodies than far away. Even on the surface of Earth time runs slower than time in deep space, but only by about 1 part in a billion.

To put that in perspective, after a hundred years, a clock in deep space and on Earth will differ by only three seconds. You shouldn't be surprised that the effect is so small, either. If it were big, it would be part of your physical intuition. However, as we will see, near the event horizon of a black hole, this effect becomes really significant. Compared to distant observers, an astronaut at rest near the event horizon will appear to be moving infinitely slowly.*

In this chapter in particular, we're introducing a lot of pretty fanciful stuff: wormholes, time machines, cosmic strings, and the like. The reason we're starting with black holes, though, is that black holes almost certainly really exist. We've nearly seen them—we think.

Before we tell you about the observational evidence for black holes, perhaps we should clear up a couple of misconceptions right off the bat.

1. Black holes are not the unstoppable killing machines they are made out to be. For instance, if our Sun were suddenly to turn into a black hole, nothing interesting would happen. Well, that's not entirely true. We'd die, of course, but only for the prosaic reason that we'd freeze from lack of sunlight. However, Earth wouldn't suddenly be sucked into the now black-hole Sun.† While the size of the object has changed, it still functions under the same rules. The gravity at a distance will stay the same, and Earth will continue to circle in the same orbit. Gravity far away from black holes behaves exactly like gravity from any other body with the same mass.

2. Black holes aren't really completely black. It's true that light can't escape, but we believe that black holes *do* radiate a small amount of light from their surfaces.

   In 1974, Stephen Hawking theorized something interesting. While nothing can escape from a black hole, the area just outside is an extremely dynamic place. Particles and antiparticles (take, for example, electrons and their evil twins, positrons) are constantly

---

*The workers of the Department of Motor Vehicles are totally dedicated to customers, and we refuse this opportunity to take a cheap shot at them.
†Which has come, presumably, to wash away the rain.

being created and annihilated in pairs, as we saw in chapter 2. Imagine a pair of particles, an electron created just inside the event horizon, and a positron, created just outside. The electron, of course, will never be seen again, but the positron might have been called into existence with just enough energy to get away. The positron eventually gets made into energy that might be observed somewhere far away. Of course, this same trick could be done with any particle/antiparticle pair, including photons, which are their own antiparticles. The upshot is that a black hole, left to its own devices, will start to give off energy and radiation.

This seems like we're getting something for nothing, but this extra energy came from the mass of the black hole. This model of Hawking radiation, as it's known, predicts that all black holes will eventually evaporate all of their mass in this fashion.

But don't hold your breath waiting for this to happen.

If you started with a black hole the mass of the Sun, it would take 10^{57} times the age of the universe for it to evaporate.

Everything we've just said comes from theory—interpreting what Einstein said about general relativity (with a bit of quantum mechanics thrown in for good measure), and making predictions about what black holes should be like. Nevertheless, there is very good evidence to suppose that black holes are real and that they come in a wide range of sizes and colors—or sizes, at least.

Some of the smallest black holes in the universe are probably not much more massive than our Sun. In our basic model of stellar evolution, welterweight stars like our Sun will use up their hydrogen in about ten billion years or so. After a stint as a red giant, and then another attempt at making something useful out of its helium, the Sun will ultimately slough off a gaseous envelope. Only a smoldering white dwarf will remain.[*]

For stars more than two or three times the mass of the Sun, something altogether different happens. These stars end their lives in an

---

[*]Try rereading the paragraph, replacing "our Sun" with "Nick Nolte." It's fun!

enormous explosion called a supernova. Most of the lighter stars end up as a very tightly bound ball called a neutron star, and some special few of the most massive end up as black holes. Astronomers have seen lots of supernova explosions, though thankfully none of them has occurred in the vicinity of Earth. This is good, as a nearby explosion would result in an untreatable case of deadness to humans. But we've never seen the remains; we've never really seen a black hole.

How, then, can we be so confident that black holes exist? While we don't see the stellar mass black holes, we do see signs of super-massive black holes at the centers of large galaxies, and nowhere is the evidence stronger than in our own.

In the mid-1990s, a number of astronomers, including Rainer Schoedel of the Max Planck Institute, and Andrea Ghez of UCLA, started observing the motions of stars in the center of our Galaxy. By 2002 their observations bore fruit, and what they saw was remarkable. Less than a light-year from the center (which is *really* close in by galaxy standards) they measured the positions of stars and found that year after year, they seemed to be moving. Indeed, the stars were almost certainly in orbit around something *very* compact and *very* dark.

Over the past few years, as the measurements have gotten better and better, and the paths of the stars were traced for longer periods of time, it became more certain that the object at the center of our Galaxy was a black hole about four million times the mass of the Sun—enormous by our standards, but nothing compared to some of the biggest black holes out there.

Many of the most distant objects we know about are powered by supermassive black holes. Even though black holes themselves give off very little light, they exert an enormous amount of gravitational pull, especially as things get close to them. As gas falls closer and closer to the central black hole, it speeds up and starts to give off lots of radiation. These black holes, surrounded by stuff, give off a huge amount of energy and are called quasars. Quasars give off so much light that they can be seen almost all the way across the universe. And at the centers of them are black holes often more than *a billion* times the mass of the Sun.

What happens if you fall into a black hole?

We started this whole discussion with one simple objective: build a time machine. It seems a little excessive, but we had to talk about black holes first—and it makes sense. Gravity warps time, and black holes are a ripping source of gravity, so maybe we can use a black hole to travel through time. Though most models for time machines aren't based on black holes, they are simple enough that we can get some sort of gut feeling about how the bending of time actually works before we delve into the nuts, bolts, and cosmic strings of chronoengineering.

So if we're going to really build a practical time machine, we need to get our hands dirty by jumping into one of these time warpers feet first. By way of example, we'd like you to consider what would happen if Dr. Dave and Robo-Jeff decided to mount an expedition into Oblivion, a ten-solar-mass black hole, and claim it for the Evil Research Academy on Jupiter.

Dr. Dave, being the more cautious (and perhaps smarter) of the two, decides to stay back and take observations back at the Evil Research Academy while Robo-Jeff (who is certainly the more handsome, in a rugged but noticeable way) sets himself up in a space suit, complete with a radio transmitter/receiver and blue racing lights.

Of course, from out in deep space, Oblivion doesn't look like much. If they could see the event horizon (which they can't) it would look like a globe about eighteen miles in radius. There are a few differences, however. Because the gravitational field of Oblivion is so strong, it actually bends light, allowing Robo-Jeff and Dr. Dave to see the stars *behind* it!

Of course, they can't sit around admiring Oblivion all day, and Robo-Jeff eventually takes the plunge, falling feet first toward Oblivion. At first he doesn't notice much of anything, and simply falls faster and faster toward the black hole. By the time he's about ninety-three million miles away (the distance that Earth is from the Sun), he'll be falling at a rate of more than three hundred thousand miles per hour.

A dizzying speed, to be sure, but since he's in free fall, he would feel weightless the whole time.

As he gets closer and closer, a curious sensation begins to take over.* The gravitational pull on his feet will be stronger than the gravity near his head. At first this seems like just a mild disorientation, but by the time he's about four thousand miles (the radius of Earth) from the center of the black hole, the *difference* between the gravity at his feet and at his head will be the same as the total gravity here on Earth. It would be akin to a crane holding him by his cranium while his feet dangle toward the ground.

This tidal force is relentless, and as Robo-Jeff gets closer and closer to the center of the black hole, he finds himself getting stretched out quite dramatically. Astronomers call this process "spaghettification." With the exception of Plasticman and Mr. Fantastic, human bodies like Robo-Jeff's don't so much stretch as *break* when subjected to normal forces. The tidal forces should be fatal, as the record for humans surviving high accelerations is about 179 times Earth gravity, but even this was only for an instant (in a crash). Robo-Jeff will experience this (and worse) continuously once he reaches about 720 miles from the center of Oblivion.

By the time he reaches a distance of about 350 miles from the center, the difference in gravity between head and feet will be about fifteen hundred times Earth normal, strong enough to literally rip apart human bones.

You *knew* time travel wasn't going to be pretty.

---

*That tingle means it's working.

Let's pretend that Robo-Jeff has been eating his Wheaties and that his bones and robotic limbs can withstand the awesome forces. Even with this generous consideration, it occurs to him that he has failed to install a rocket booster to get him out of Oblivion's gravitational field. By the time he gets about forty miles from the center, he finally starts to panic (in a manly, consequences-be-damned kind of way) and sends an SOS call to Dr. Dave at the Evil Research Academy. However, because the photons from Robo-Jeff's radio transmitter lose energy as they travel outward, Dr. Dave has to adjust his dial to a much lower frequency to hear Robo-Jeff's cries for help.

As Dr. Dave listens on his radio, he finds that even though Robo-Jeff is transmitting at 108 MHz as agreed upon ahead of time, he only hears him at the very bottom of the dial, in the NPR territory. This is the exact phenomenon discussed earlier; the photons that Robo-Jeff sent (in the form of a radio signal) have lost energy and so appear farther down the dial. When Dr. Dave finally tunes him in, Robo-Jeff's voice sounds slow and deep, like when you play a 78 on the wrong speed setting, or Barry White on the correct setting.

As Robo-Jeff continues falling, he loses radio contact entirely.

Although the pair failed to install a rocket booster, they did think far enough ahead to install blue racing lights on his suit, and Dr. Dave continues watching Robo-Jeff by these same flashing lights. Rather than gleaming blue, they seem greenish, and then yellowish, and then red before becoming invisible to the naked eye. At that point, Dr. Dave can only observe Robo-Jeff using his infrared detector.

About eighteen miles from the center of the black hole is the event horizon, and as time goes on, Dr. Dave notices that while Robo-Jeff is getting closer and closer to the point of no return, he never actually crosses it. During the entire fall, Robo-Jeff appears to be *outside* the black hole. However, the lights on his suit eventually just get redshifted out of Dr. Dave's detector range and he seems to disappear.

From Robo-Jeff's perspective, on the other hand, everything seems to happen in fast forward, and the signals from the Evil Research Academy appear high-pitched. And what happens at the instant he crosses the event horizon?

Except for the fact that he will most likely not be among the living and that he can no longer escape, he wouldn't notice anything special. Robo-Jeff might not even notice that he crossed the event horizon. He will just keep falling inexorably toward the singularity. Of course, once he's on the unhappy side of the event horizon, photons can no longer move outward, and therefore there is no scenario in which he is not ripped completely to shreds. It might be of some comfort to him that the entire period from the instant when he started to be uncomfortable (when he felt tidal forces of about 10 g's) to his complete destruction turns out to be only about a tenth of a second.

This, by all scientific evidence, still sucks for him.

Can you go back in time and buy stock in Microsoft?

As we've just seen, the regions around black holes have gravitational fields that severely warp space, and more importantly for our nefarious purposes, time. The big question is whether Dr. Dave and Robo-Jeff will be able to use general relativity to build the greatest piece of mad science ever: a time machine. Before we start talking about *how* to build a time machine, we should offer some idea of what a *good* time machine might be like.

When we were children, we liked to play with refrigerator boxes, and occasionally would write the phrase "Time Machine"* on the side of one of them. By a very generous interpretation, it *was* a time machine. After all, it allowed the occupant to move through time at a rate of one second per second. We imagine that you're hoping for something a little more flexible than that.

Ask and ye shall receive; we *can* do better. As we've seen by Robo-Jeff's bold example, standing near a black hole or a white dwarf slows the personal clock and therefore allows travel through time at *faster than* one second per second. Our evil duo could make a pretty decent time

---

*Ideally with one or two adorable backward letters on it, in the tradition of Toys "Я" Us.

machine to the future by exploiting this. For example, they could build a ship to plunge down to right outside the event horizon of a black hole, hover there for a short while, and then come back out, producing a return time well into the future.

This is a one-way trip, however, since it allows them no way to go back to their original time. What we really want is to send them into the past, and in the best-case scenario, change it to suit their own nefarious plans.

So what are the prospects of going into the past? As we saw in chapter 1, we can certainly arrange to *see* into the past. When you look at *anything*, you're seeing it as it once was.

Of course, you might have something more specific in mind. For example, imagine you wanted to observe something particular, such as the Crimean War or the Apollo Moon landing. In principle, this seems pretty easy. For the Moon landing, you'd just need to park a spaceship forty light-years from the Moon with an ultrapowerful telescope.* The problem is that to get forty light-years from the Moon, it would take at least forty years to get there, since nothing can travel faster than the speed of light. So while we can see into the past, we can't generally see into our own past history, since we can't outrun light without using gravity to cheat.

Of course, a mirror is a handy work-around. If there happened to be a mirror about twenty light-years from the Moon already in place, then in principle the Moon landing would just be reflected to us now. Unfortunately, since the mirror would already have to be in place, we'd have to be *very* lucky. Also, the picture would be very small.

While even seeing into the past has its limitations, to most people a time machine should actually allow you not only to see into the past, but also to interact with and potentially *change* it. At the very least, you'd like to be able to go back in time and shake your own hand.

General relativists refer to scenarios in which you could encounter yourself (or, in principle, your ancestors) as a "closed timelike curve." As we will see shortly, there are candidate designs for time machines

---

*As of this writing, it's been about forty years. Also, the numbers come out easy.

that are perfectly consistent with relativity. They might even allow for the real possibility of meeting your younger self.

But before we do that, we need to lay down a few ground rules.

We fully recognize that at this point in the conversation we're outside the realm of what might properly be considered physics, and well into philosophy land. And we have no problem with that. In both fiction and scientific philosophy, there seem to be two basic pictures of how time travel might play out:

1. Alternate Realities/Universes

One of the most obvious problems with time travel is that it gives you apparent license to muck about in the past in pretty much any way you like. For example, imagine you got it in your head to do something profoundly stupid like kill your own grandfather before your father was conceived.* Could you do it? What would happen to you afterward? What kind of future would you find when you returned to the present?

By killing your grandfather, you yourself couldn't have existed, and therefore it seems impossible that you could have gone back in time in the first place, and therefore you couldn't have killed your grandfather, and so on.

How can we resolve this "grandfather paradox"?

A possible explanation for drastically changed pasts comes to us via quantum mechanics. In chapter 2 we discussed Hugh Everett's Many Worlds interpretation of quantum mechanics in which every quantum event gives rise to parallel universes. As we've seen, at the microscopic level, the universe is *really* random, and no amount of knowledge could predict, for example, if a radioactive atom will or will not decay within a particular time period, or whether a particular electron was found to be spin-up or spin-down. If we replayed the movie of the universe again, would the same thing happen again? We have no way of knowing.

While a small thing such as the spin of an electron might not seem like much, taken over a very long period of time, small things can

---

*We don't know why physicists have gotten into the macabre habit of talking about grand-patricide, but who are we to argue?

add up to a lot. Remember the old proverb, sometimes attributed to Benjamin Franklin:

> For want of a nail the shoe was lost.
> For want of a shoe the horse was lost.
> For want of a horse the rider was lost.
> For want of a rider the battle was lost.
> For want of a battle the kingdom was lost.
> And all for the want of a horseshoe nail.

This saying is just the conceptual version of a theory in mathematics called chaos. It's true for almost any system (and the history of humanity is no different) that even a small difference in the starting point can cause enormous differences in the final outcome. You may also know this as the butterfly effect,* in which as small a thing as the flapping of a butterfly's wings can change the weather months later on the other side of the world.

The point is that while each of the parallel universes may start almost identically, they diverge into vastly different histories in short order.

This same picture—that of universes branching off one another—could also be used in models of time travel. Let's see how this resolves the grandfather paradox. Imagine you're a time traveler living in universe A, and you get it in your head to screw up the universe in some spectacular fashion. You build a time machine, go back to the past, and kill your grandfather. Since that event simply didn't happen in the history of universe A, the murder must occur in a new universe, B. If we then went forward in time, presumably we would find ourselves as us (with our own memories), but in universe B, rather than in our original universe. And, of course, there would be only one of us (version A) in universe B. The other one was never born.

See how simple time travel logic can be?

The Many Worlds model is the basic one used in *Back to the Future*. This movie is such a classic that you're probably already familiar with it. A teenager, Marty, uses a souped-up DeLorean to go thirty years back in time,

---

*Not to be confused with the terrible "time travel" movie of the same name.

inadvertently complicates the courtship of his parents, and spends the rest of the movie trying to undo his mistakes and get back to his own time.

Of course, he succeeds. But when he gets back to his own time, the history of the world has changed significantly. Put into the Many Worlds picture, Marty A disappeared from universe A to go back in time. When changing the future, he branched off into universe B and went back into the future of universe B. Meanwhile, Marty B presumably disappeared into the past, changed the timeline, went back into universe C, and so on.

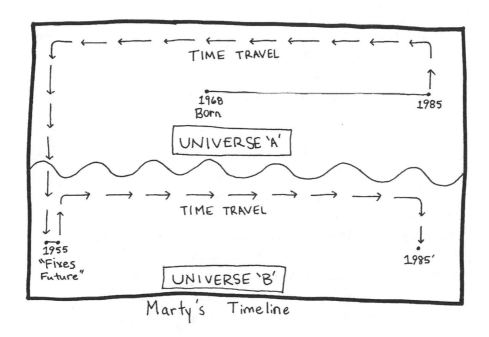

Marty's Timeline

In principle, if Marty A changed the universe enough, Doc Brown B (the inventor of the time machine) never would have invented time travel. In 1985, Marty B would then be stuck in universe B, unable to travel to the past. When Marty A returned to the present, he would do so in universe B, and so there would be *two* of Marty. Meanwhile, in universe A, Marty would disappear with the DeLorean, never to return.

In the Many Worlds model, even if you somehow resisted the urge to kill your ancestors or otherwise mess up your history in an obvious way, we still would have to worry about how your past actions affected the future. Even something innocuous rerandomizes all the events of the past.

The butterfly effect pretty much assures us that there is no time travel scenario in which you can really change the past *without* introducing parallel universes. To us, though, this explanation is just a cheat. A non-time-traveling observer in that universe would see duplicate people, dead grandfathers, and time travelers popping into and out of their reality. To us, this is deeply unsatisfying.

2. The Universe Is Self-Consistent

The thing that really distinguishes physics from magic is that physics makes testable predictions about the universe. Thus far there's no direct experimental evidence (nor any real proposal of how to get any) indicating that ours isn't the only universe there is. If there's only one universe, then there's only one version of history.

In the mid-1980s, Igor Novikov of the University of Moscow developed a theory in quantum mechanics and time travel that says that the probability of a self-inconsistent history is identically zero, but be aware that this theory is premised on the assumption that parallel universes don't exist; if we put in parallel universes, then all bets are off.

In the self-consistent model a realistic time-travel scenario might go as follows. At age eighteen, an older version of yourself comes back and gives you general instructions on how to build a time machine. Realizing your destiny, you spend the next ten years building a time machine, then go back in time to subsequently give yourself the same instructions.

But here's the dilemma. What if you tried to kill yourself in the past? Or simply didn't give yourself instructions on how to build the machine? Is it even possible for you to do something that removes the motivation for you to build the time machine in the first place?

Time travel gives us two unsavory alternatives. If you subscribe to the Many Worlds model of time travel, then Novikov's theory is violated. On the other hand, in the self-consistent universe model, a time traveler apparently doesn't have free will.

We (that is to say, the physics community) don't have any particularly good answers to this quandary. We simply assume that the physical laws require that however time travel works, self-consistency must be maintained.*

Self-consistent universes are much tougher, both for writers and for the real universe. In part, you have to ask the question of why you would go back in time in the first place. You never could have a motivation to, since going into the past couldn't fix anything. On the other hand, if you wanted to engage in a bit of chronotourism, presumably nothing would prevent you from observing the fall of Rome or the first Olympic games. Of course, an astute observer at the time presumably would already have seen you in the audience.

We're going to take as our starting point that "good" time travel narratives necessarily involve the self-consistent history model. For one thing, it's much harder to make a riveting self-consistent history using time travel, and we feel that people should be rewarded for doing so. For another, since there's no evidence for parallel universes, the single-history version of time travel is the one most in keeping with what we actually know about physics. Mostly, we simply don't like stories that involve parallel universes because even if something is "fixed" in one universe, it usually remains broken in another. It's fine for you to fix your own timeline, but if that means that countless other universes become dystopian nightmares, is that a risk you're willing to take?

 ## Who does time travel right?

How does popular entertainment fare by this standard? Books, by and large, do pretty well at keeping things self-consistent. Some, such as the classic *The Time Machine*, avoid consistency altogether by spinning a narrative that takes place in a distant enough future that it's pretty clear that the time traveler couldn't change the future even if he wanted to. Others, such as Douglas Adams's *Hitchhiker's Guide* series, are so clearly absurd that they're not meant to be literal time travel stories at all.

---

*Yes, we realize that this assumption is a bit of a cop-out. If you really expected us to reconcile the issue of free will versus determinism, then you expected way too much for the purchase price of this book.

Movies and television do much worse, typically. Most of them (*Back to the Future* or the television series *Heroes*, among the most obvious examples) take as their starting point that the future is not yet written. Nonsense! Of course it's been written if you've actually been there! Your entire motivation for taking action in the present (or past) is predicated on the fact that you've seen what the future looks like.

Being science-fiction geeks (as well as regular-science geeks), we can't but look for errors when a movie or a TV show uses time travel as a plot device. Sometimes, though, they get it just right. With that in mind, Robo-Jeff has kindly compiled a (noncomprehensive) "Two-Sentence Time-Travel Summary" at the end of this chapter. Meanwhile, a few detailed case studies are in order. Before we proceed, we should warn anybody who's managed to avoid TV and the movies over past thirty years or so that there are severe spoilers ahead.

Futurama Season 4, Episode 1, "Roswell That Ends Well" (2001)

A thousand years in the future, technology will be well beyond what it is today, and people will be able to travel back in time (unpredictably, but nonetheless effectively) by microwaving metal while simultaneously watching a supernova.

The *Futurama* crew, which includes Philip Fry, a delivery boy cryogenically frozen for a thousand years, and Bender, a devious bending robot, travel from the year 3001 to 1947 Roswell, New Mexico. Upon landing— crashing, actually—Bender's head and body are separated, and Fry keeps the head. Bender's body, on the other hand, is mistaken for a flying saucer— the very UFO covered up by our government in our own history.

Fry discovers that his grandfather is on the local army base and accidentally kills him. While comforting his grandmother, Fry comes to a realization: since he still exists, his grandmother *can't* be his grandmother!

The next morning, Fry comes to another, creepier realization: the woman *is* his grandmother, and he unknowingly became his own grandfather.* Since interference in a self-continuous time loop is impossible, it is better to say that he has *always been* his own grandfather, and he

---

*The show, thankfully, glosses over the relevant details.

has merely fulfilled his obligation to the timeline. Not that that excuses his actions.

During their escape from Roswell, Bender's head falls out of the ship, and the crew is forced to escape back to their own time in the thirty-first century without it. Fry realizes that it must still be in the desert (and indeed, has been there all along), and the crew digs it up and reattaches it to Bender's body.

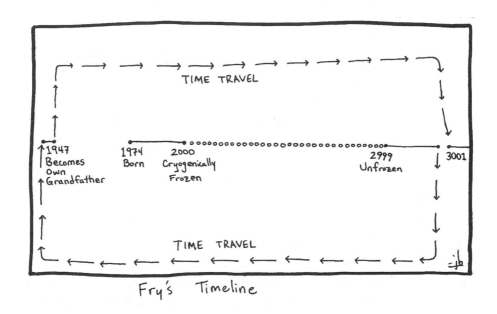

Fry's Timeline

This plot line introduces some complications into the lives of Fry and Bender: particularly, Fry is his own ancestor, and Bender possesses a head that is more than a thousand years older than his body. While these things seem spectacular and silly, there is no scientific reason why they cannot both be true.

The Terminator (1985)*

One of the most dramatic and anticipated events of the future is the nuclear robot holocaust, a festival of burned carcasses and sparking

---

*The original *Terminator* movie is a paragon of time-travel awesomeness. The sequels, not so much.

metal chassis, where cyborgs equipped with arm-mounted chain guns rend the flesh of the living, care of Skynet©, a psychotic and sentient version of our present-day Internet.

And we couldn't be more excited.

Particularly, we look forward to the semirealistic time travel that will inevitably occur. In the future, John Connor is leading the rebellion against an army of evil, murderous robots. John sends a soldier, Kyle, to the past (our present, or at least, 1984) to protect his mother, Sarah Connor, after Skynet© sends one of its own agents, a murderous robot assassin,* to kill her.

Kyle, working from a photograph for identification, finds Sarah Connor and does his best to keep her safe. He is infatuated with Sarah, falls in love, and impregnates her—a baby who grows up to be John Connor, the leader of the rebellion against the machines.

Not only do Kyle and Sarah neutralize the robot threat and save Sarah (and John's) life, but Sarah also gets her picture taken, a token that will eventually be passed to Kyle, who will fall in love with her all over again. The loop is self-consistent, and if Skynet© had only thought about the conundrum for a few minutes, it would have realized that since John Connor wasn't dead in the future, he *couldn't* have been killed in the past, and that the entire attempt was futile in the first place.

Of course, if Skynet© had realized the futility of its mission, then it wouldn't have sent a Terminator back in time to kill Sarah Conner, Kyle wouldn't have followed the robot into the past, and John wouldn't have been born. Whoa!

Does this mean that mankind will have to battle enormous odds to end the robot insurrection at some point in the future, led by a computer network that can't fathom self-consistent time loops? We guess: yes.

 ## How can I build a practical time machine?

We've already established that general relativity can do wonky things to the flow of time, and we've even set some ground rules regarding what a time machine should and shouldn't be able to do. It may interest you

---

*Currently serving as governor of California.

to know that there are real physicists out there who are busily publishing papers on whether and how a practical time machine can be built.* Having set up all the ground rules, we're finally going to get around to the central question of how we might go about building a time machine that's consistent with what we know about physics.

1. Wormholes

General relativity shows that massive bodies such as the Sun or a black hole will warp space and time. However, the warping of space is a *local* phenomenon. We mean that if you take a flat sheet of paper (flat = not warped) and roll it into a tube, a tiny ant wandering on its surface couldn't tell whether it was rolled.

In principle, we could exploit the fact that space could be "folded" in order to build a time machine. This fact is central to the idea of a wormhole—a staple of science fiction for decades. A wormhole is a theoretical solution to Einstein's equations of general relativity in which space gets so distorted that a path is created connecting two potentially distant regions of space.

HOW A PHYSICIST MAY EFFECTIVELY CHEAT IN A GAME OF TAG

---

*Rest assured, they already have tenure.

Far away, a mouth of a wormhole looks kind of like a black hole. From our local perspective, it would be a sphere through which you could see out the other mouth of the wormhole. However, unlike a black hole, as you get closer and closer to a wormhole, gravity stops getting stronger, and a person or spaceship could go through without being ripped apart.

While we have serious indirect evidence that there are black holes out there, there is no evidence, either direct or indirect, that wormholes really exist, and suspicions are that, Arthur C. Clarke's hoping and praying aside, they probably don't exist on a macroscopic scale. All we know from general relativity is that they *could* exist.

The basic picture is that you get in one end of the wormhole and come out the other, far away. In fact, you could design it so you could easily travel faster than light. We'll forget about the difficulty in building such a thing for the moment and point out the obvious. While wormholes sound like great teleportation devices, it's not obvious how you'd use them as time machines. This is fine, though, because Kip Thorne of Caltech has done the heavy lifting for you. In his book *Black Holes and Time Warps*, he describes a wormhole time machine design that he and two of his students, Michael Morris and Ulvi Yrtsever, proposed in 1988.

To transform an awesome teleportation device into a super-awesome time machine, you first need to realize that the length of the inside of the wormhole has no relation to how far you can travel using it. If you were to go through a wormhole, you'd come out (from your perspective) a very short time later.

Let's make things concrete. Earlier, we told you about prudent (and stuffy) Dr. Dave and adventurous (and foolhardy) Robo-Jeff and their adventures exploring black holes. Well, they're at it again, only this time they've managed to build themselves a small wormhole—big enough for a person to go through, but small enough to put one of the "mouths" inside a spaceship, which is exactly what they do. If Dr. Dave looked through one mouth of the wormhole (placed, conveniently, in his living room where the TV normally might go), he would be able to see into the interior of Robo-Jeff's ship.

On January 1, 3000, Robo-Jeff takes his ship and wormhole and flies off at 99% of the speed of light. He travels about seven light-years from Earth and returns on January 1, 3014. If these numbers seem a bit familiar to you, they should. We picked the exact same ones when we talked about the twin paradox in chapter 1.

You will also recall that from Robo-Jeff's perspective, only two years will have passed. Here's where things get weird. Dr. Dave and Robo-Jeff can see each other through the wormhole. The interior of the wormhole doesn't know that anybody's moving. So if Dr. Dave spends the next two years of his life watching Robo-Jeff through the wormhole, he will see Robo-Jeff make his preparations for blastoff, pilot his ship for a year, turn around, and return home. Dr. Dave will fully expect to go out on his lawn in 3002 and find Robo-Jeff waiting for him.

He will be disappointed, and will stare at the sky morosely for the next twelve years until Robo-Jeff returns to Earth with the other side of the wormhole.

So consider this. Looking through the wormhole in his living room in 3002, Dr. Dave *sees* Robo-Jeff land on Earth in 3014. He can literally see the future. But it's even better than that. He can visit the future, or, for that matter, Robo-Jeff can visit the past. Actually, anybody else can as well. The wormhole now becomes a way to travel twelve years in the past, and also to travel the negligible distance from Dr. Dave's lawn to his living room.

But beware! Though this time machine would allow you to travel in the past, you couldn't simply do anything you liked in the past, for the reasons we already discussed. After all, the past has already occurred.

There is also another severe limitation. You can't go back in time to *before* the time machine was built. This might help answer the nagging question that has probably already occurred to you: why aren't we visited by time-traveling tourists? Because we haven't built any time machines yet!

There are other problems with this design. For example, it is very hard to hold a wormhole open, since the natural tendency is for it to pinch off whenever matter or energy passes through it (since gravity will attract the sides of the wormhole). A wormhole might collapse before it could be used for anything useful. To hold it open, Thorne asserted that it needed to be held open by some "Exotic Matter," which has a negative energy density. While this doesn't seem to be in abundance (or even existence) under normal circumstances, the same sorts of fields that cause black holes to radiate have exactly the properties we're looking for.

And even that may not be enough. One of the problems with the wormhole model is that it combines two areas of physics we haven't yet successfully merged: quantum mechanics and general relativity.

Our verdict: good luck making a wormhole time machine. Wormholes may exist on the microscopic scale, and they may not, but so far there's no indication that there are spaceship-size wormholes out there, nor do we have any idea how to make them. Once made, there's every possibility that your wormhole will collapse before you, or anything else, can traverse it.

2. Cosmic Strings

Cosmic strings have little (or no) relation to the strings of string theory, except they are also based on analogy to the everyday string that kittens play with. They are exceedingly dense, and either infinitely long or folded into a loop. As you might imagine, they produce enormous gravitational fields, and because of that, they warp space enormously.

In 1991, Richard Gott of Princeton University developed a time machine model based on cosmic strings, and he gives an excellent description of his model in *Time Travel in Einstein's Universe*.

In general relativity, it's not always true that the shortest path between two points is a straight line. We can exploit this fact to do all sorts of interesting things involving apparent "faster than light" travel. For example, imagine that we have two cosmic strings lined up halfway between Earth and a distant planet, Quagnar VII.

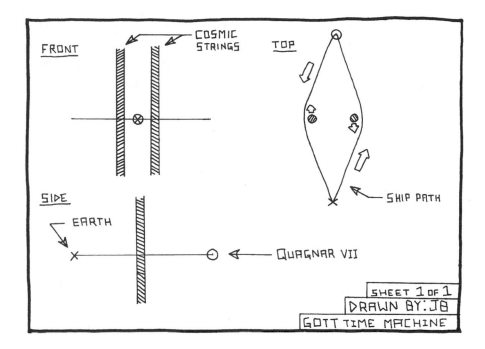

Robo-Jeff decides that he wants to travel to Quagnar VII as quickly as possible. Because cosmic strings warp space and time around them, it's faster to go around the strings than straight through the middle. If a laser were fired straight down the center at the same moment that Robo-Jeff took off, his ship could beat the light beam even though he can *only* reach speeds of 99.9999% the speed of light.

This last point is important, because light beams are the be-all and end-all of relativity. Think about what happens if Robo-Jeff's kid brother, Robo-Dan, launches from Earth to Quagnar VII at very high speed but takes the middle path. He is astonished to note that Robo-Jeff arrives at Quagnar VII before the light beam. In fact, by his reckoning, it's possible that Robo-Jeff arrived *before* he (and hence the light beam) was launched from Earth. We suppose this is a sort of time travel, but not a particularly useful sort. Even though Robo-Dan says that Robo-Jeff arrived before he left, he couldn't do anything useful with that. Robo-Jeff couldn't, for example, go back and shake hands with his past self, because by the time he got back, he would have long since left. Got it?

We can turn our cosmic strings from a curiosity to a practical time machine by setting them in motion at nearly the speed of light. To make things relatively simple, imagine the string on the right moving toward Earth and the string on the left moving toward Quagnar VII, each at the same very high speed.

We use the same trick that we employed when talking about the wormhole time machine. Dr. Dave sits at the midpoint between the two cosmic strings, and because he's not moving, his clock moves at the same rate as an observer on Earth.

Now here's the cool part. Robo-Jeff leaves Earth and travels a counterclockwise loop around the two strings. We've already found that to an observer moving through the middle of the cosmic strings, Robo-Jeff appears to arrive *before* he departed.

It gets better. On the return trip, Dr. Dave sees exactly the same thing, except that Robo-Jeff flies around the leftward string. Again, Robo-Jeff appears to arrive before he left from Quagnar VII, which, in turn, is before he left Earth.

Let's say it again: Robo-Jeff arrives back on Earth before he left according to both Dr. Dave and, more importantly, the people on Earth. In this scenario, he could go back in time, shake hands with himself before he left, and alter history in whatever way the laws of time travel will allow.

Of course, there are still some important caveats. Like with the wormhole time machine, he *still* couldn't go back to before the time machine was built.

There's also an important physical conundrum. There's no observational evidence that cosmic strings exist, and if they don't, they'd be difficult (if not impossible) to create. For one thing, this particular design requires cosmic strings that are infinite in length, and it would take an infinite amount of time to create one. There's also the very real problem of accelerating giant strings to nearly the speed of light.

Our verdict: while we don't like to throw the word "impossible" around, let's just say that a cosmic string time machine would certainly be a challenge.

What are my prospects for changing the past?

At the end of the day, can you build a time machine?

You? Almost certainly not.

Is it physically possible for a super-civilization? Perhaps, but this is strongly dependent on the existence of things such as wormholes, exotic matter, or cosmic strings, as well as the technology for the harnessing and manipulation of enormous energies.

However, there are some very real constraints. Every practical design for time machines using general relativity has had two built-in safety mechanisms. First, the time machines would only let you visit periods after the machines were invented. Second, and perhaps more important, all of them are consistent with Novikov's theorem that the universe has just a single version of history.

In response to Thorne's wormhole time machine, Joe Polchinski, then at the University of Texas, raised the issue of whether we could build an experiment that was the equivalent of the grandfather paradox, but with pool balls. To set the stage, we put a wormhole in a rocket ship, but we create a time difference of only about three or four seconds instead of twelve years.

Imagine you shoot a cue ball into one mouth of our wormhole time machine. If it's the "later" mouth, then some time before you sink the shot (though potentially after you've already struck the ball) a second ball will come flying out of the "earlier" mouth.

Think of this as a variant on a minigolf hole in which you hit a ball into a cup at the top of a hill and it comes flying out through a pipe at the bottom of a hill, except that in this case, you manage to arrange things so the ball comes flying out of the second hole *earlier than* when you shot it into the first.

Presumably a skilled enough shooter could hit a pool ball in the first pocket so it comes flying out of the second (earlier) end of a wormhole just in time to mess up your original shot. But if your shot gets messed up, then what came out to disrupt your shot?

Don't try this shot, because it's not going to work the way you think it will.

Thorne and his students studied the problem using the tools of quantum mechanics. Remember that in chapter 2 we saw that according to quantum mechanics, a particle takes all possible paths to go from point A to point B, and that different possible paths can interfere with one another, producing a single observed result. The same basic thing can happen in our time machine, forcing the earlier and later versions of the pool balls to interact only in a way that is entirely consistent with a single history.

Imagine you tried to make the trick shot we just described. What would end up happening (according to your view) is that you'd take your shot, but before your ball went in the first (later) end of the wormhole, an identical-looking ball would fly out of the second (earlier) end, knocking your ball slightly. Your ball would still go into the wormhole, but at a slightly different angle than you'd intended. Remember that you lined up your shot with the sole intention of blocking your own shot, and now that's been screwed up. In fact, at the angle your shot

went in, you'd expect the ball to come out of the second (earlier) end of the wormhole at exactly the angle it *did* come out. And so it did.

In other words, travel through time all you want. The present will still be waiting for you when you return.

Robo-Jeff's Two-Sentence Time-Travel Summary

Important note: We're not attesting to overall quality here, just how well these movies and TV shows do with a self-consistent or alternate universe time travel model.

12 Monkeys (1995) ★★★★★   A Philadelphia-based super mystery that shows why you should never share living space with Brad Pitt. The film also offers compelling evidence against watching future versions of yourself commit suicide.

Back to the Future (I, II, III; 1985, 1989, 1990) ★   Changing your past does not make you slowly disappear. Sorry, Oedipus.

Conquest of the Planet of the Apes (1972) ★★★★★   A super-intelligent chimp, descended from future apes, goes back to 1991 and leads the simian revolt. Darwin would be annoyed to discover that it takes only five years for gorillas to learn to speak fluent English.

Heroes (TV; 2006–) ★   Self-aggrandizement reaches new heights when Hiro finds out that the person he's worshipped all his life is himself. He later goes on to undo a future that is already written and must never happen.

Primer (2004) ★★★★   Two best friends accidentally build a time machine out of argon and old mufflers. Eventually they (or their doubles) kill their doubles (or themselves), but other more different doubles (or themselves) go through time trying to stop themselves (or something).

Quantum Leap (TV; 1989–1994) ★   Dr. Samuel Beckett and his imaginary friend invade the past, and other people's bodies, to correct the timeline. If the show is to be trusted, a scientist teaches Michael Jackson how to dance.

Star Trek IV (1986) ★★★★   Who knows whether Kirk and his crew changed the past? All they did was kidnap some whales.

The Time Machine (1960) ★★★★   George Wells travels eight hundred thousand years forward in time to discover a symbolic bifurcation of good and evil. No past was altered in the making of this future.

Timecop (1994) (no stars)   In 2004, time travel is illegal. Jean-Claude Van Damme is (predictably) a time cop who saves his (supposedly) dead wife's life without changing the timeline.

The Expanding Universe

"If the universe is expanding,
what's it expanding into?"

We have to give credit where credit is due. Thanks to write-ups in the *New York Times*, specials on PBS and the Discovery Channel, and other popular books on this subject,[*] certain scientific phrases have crept into the public consciousness. For example, ask someone on the street what's happening in the universe at this very moment, and odds are they'll tell you that the universe is expanding. Go ahead; we'll wait.

Now go back and ask that same person what it means exactly for a universe to be expanding. We bet that he or she is not going to give such a pat answer this time around. That's where we come in.

First a word on what it doesn't mean. Do you remember the scene in *Citizen Kane* in which Charles and Emily are at the breakfast table, and over a series of years, we see the table expand and expand and the distance between Kane and his wife grow larger and larger?[†] That is not what happens when the universe expands. Your table doesn't expand. The Earth doesn't expand. The solar system doesn't expand.

---

[*]Which almost *never* have cartoons, by the way.

[†]If not, please take this opportunity to watch it. It's nearly universally agreed to be the finest American film ever made.

The Milky Way Galaxy (which is tens of thousands of light-years across) is still far too "local" to participate in the expansion of the universe as a whole.

Even the Andromeda Galaxy, about 2.2 million light-years away, is a falling toward us at a speed of about 275,000 mph, and may collide with the Milky Way, something you'd have to wait about 3 billion years to see. Douglas Adams almost certainly got it right in the *Hitchhiker's Guide to the Galaxy* when he wrote, "Space is big. Really big. You just won't believe how vastly hugely mind bogglingly big it is. I mean you may think it's a long way down the road to the chemist, but that's just peanuts to space." Much of this chapter will focus on just how empty space really is, but to give you some idea from the outset, when the time comes and the Milky Way and Andromeda start to tango, it's unlikely that any two stars will collide with each other. In all likelihood, it's not a stellar collision that will spell doom for humanity. We're afraid you'd just have to wait another couple of billion years after that until our Sun turns into a red giant and fries all life on Earth.

But let's not dwell on what will kill whom. This is a happy book, and we want to tell you about the seemingly innocuous expansion of the universe. When we look out thirty million light-years or so, virtually every galaxy is moving away from us. Stranger still, the farther away the galaxies are from us, the more rapidly they seem to be moving away.*

Vesto Slipher of Lowell Observatory first recorded the nearly universal recession of galaxies in 1917, but he had no way of knowing how far away the galaxies were from our own. Indeed, at the time there was enormous debate whether the dim smudges of light seen in telescopes were nebulae within the Milky Way, or entire "island universes" of their own. As it happens, it's the latter.

Distances to galaxies are tougher to measure than you might suppose. Despite what science fiction might tell you, we cannot fly to other galaxies or even nearby stars with a tape measure trailing behind. So when astronomers say, "It's twenty-three million light-years to the

---

*As you'll see later, that "seem to" is an important bit of legal mumbo jumbo.

Whirlpool Galaxy" (for example), you might wonder where they get their information.

As stars and galaxies get farther and farther away, they look dimmer and dimmer. We can use this effect to our advantage, using a "standard candle." Imagine you go to the store and get a hundred-watt lightbulb, screw it in, turn it on, and then walk away. As you get farther and farther away, the light will appear dimmer and dimmer. You know how bright the light is when you're close to it, so as you get farther away, you can estimate how far away you are by measuring how dim the bulb appears to you. The difficulty is that since we don't buy galaxies at Home Depot, it's hard to know how many watts they put out.

Even Edwin Hubble, perhaps the greatest observational astronomer of the early twentieth century, wasn't able to calibrate distances particularly accurately. In 1929, he had calibrated the distances and apparent recession speeds of other galaxies, measuring what has come to be known as "Hubble's Law." In his original paper, Hubble underestimated the distances to galaxies by a factor of eight, and as recently as the past twenty or so years, lots of papers claimed that distances, and thus Hubble's constant, were uncertain to a factor of two.* With data from the Hipparcos satellite (launched in 1989) and the appropriately named Hubble Space Telescope (launched in 1990), astronomers have measured the Hubble constant to within a few percent.

The other piece of the expanding-universe puzzle involves measuring how fast galaxies appear to be receding from us. This can be measured more or less the same way the police can figure out how fast you're moving on the road, using the Doppler shift. You may have noticed this effect with sound when a fire truck passes by. As the truck is coming toward you, the siren appears to be at a higher pitch than normal. As it goes away, it appears to be at a lower pitch. With light, the effect is similar, except if the source moves toward you, the light appears to be slightly bluer than normal; if it's moving

---

*Subsequently there were lots of people in other branches of physics who made fun of cosmologists (and continue to do so) because of our feeble attempts at precision.

away, slightly redder. The faster the source moves away, the bigger the redshift.

"275,000 mph?! This is it, Junior! We got this galaxy right where we want 'im!"

Suppose we were to take *Sesame Street*'s Cookie Monster and shoot him away from the Earth at 25% of the speed of light. His fur, which is dark blue, would appear in our telescopes as bright red. Through the eyes of an observational astronomer he would look like Elmo, but still wouldn't be nearly as susceptible to tickling.

We know that most books on the topic like to say that galaxies are moving away from us and leave it at that, but we have more faith in you than that. We're going to tell you what's really going on.

The universe is growing and the galaxies are, for the most part, sitting still while space stretches around them. It might seem like a fair bit of nit-picking, but it's an important point.

When some distant galaxy emits light, the photons make the long journey from their parent galaxy to us. As they do so, the universe expands, and the longer it takes the photons to get here, the more the universe expands in the intervening time. This expansion affects the light—an effect you've already seen. When a photon "expands," it really means that the wavelength of the light increases. The wavelength of light determines its color. So if the universe expands as the photon is traveling, the photon will get redder. If the source is farther away, the universe will expand more during the travel time, and the photon will have an even larger redshift.

 ## Where is the center of the universe?

If you're anything like us, you were raised to think that you were the center of the universe, and at first blush, Hubble's observations of the universe seem to confirm that theory. All of the galaxies seem to be rushing away from us (or the universe is expanding around us, or whatever), and it's hard to get past the idea that we're somehow special. After all, if all galaxies are receding from us, don't we have to be at the center?

Meet the Tentaculans, a race of astronomers from a galaxy about a billion light-years from our own. One of the leading astronomers is a creature named Dr. Snuggles. Would you like to meet Dr. Snuggles? Well, we have some bad news for you. Since his galaxy is about a billion light-years away, even if we radioed Tentaculus VII and asked for Dr. Snuggles, the original Dr. Snuggles wouldn't be in any shape to answer. If you were lucky, you might get a response from his great-great-great (times fifty million or so) granddaughter, and by the time she responded to tell us of our mistake, it wouldn't be for at least another billion years after that (assuming she gets back to us right away), by which point our descendants would almost certainly have forgotten what we called about in the first place. We can't actually meet with Dr. Snuggles, so we can't ask him what he sees when he looks through his telescope.

In fact, it's even more complicated than that, because the universe is expanding. If we send our signal to Tentaculus VII, it would take more than a billion years to place our call, and even longer to get a response. It's like trying to measure an eel. It gets all wiggly while you hold a ruler, and by the time you measure where its head is, you realize that the end of the ruler isn't where you left it.

No matter; we still know what Dr. Snuggles will see when he looks through his telescope. He will observe exactly the same thing we do here on Earth: nearly all of the galaxies in the sky seem to be flying away from Tentaculus VII, and the farther away the galaxies are, the faster they seem to be receding. The more jingoistic elements of Tentaculus have decided to interpret these observations as irrefutable proof that Tentaculus is at the center of the universe.

How can both Dr. Hubble and Dr. Snuggles be right? How can *both* galaxies rest at the center of the universe?

Imagine you are cooking a batch of blueberry pancakes. We choose this flavor for two reasons: first, they are delicious, and second, blueberries, much like galaxies, don't themselves expand during the cooking process. As the pancakes cook and the dough rises and expands, the blueberries start moving away from one another. If blueberries had sentience, every one of them would note, "All of the other blueberries are moving away, with distant blueberries moving away faster than nearby ones."*

This brings up a pretty subtle point, one that might seem kind of familiar if you look at chapter 1. Since everyone in the universe is under the impression that everybody else is flying away from them, how can we say that anybody is moving at all?

There is a general theme of nonspecialness that recurs throughout the history of science. Nicolaus Copernicus (in honor of whose insight the Copernican principle is now known) showed that the Earth isn't at the center of the solar system. In 1918, Harlow Shapley of Harvard showed that our solar system is nowhere near the center of our Milky Way Galaxy, despite the assumption otherwise. Now Hubble (and Dr. Snuggles on his own planet) has shown that our Galaxy isn't even at the center of the universe!

But as we said, no one can definitely claim to be. As an analogy, picture yourself as an ant living on a balloon. As the balloon inflates, you see all of the other ants get farther and farther away from you.

An astute nitpicker might object to the ant world. He or she might say, "Wait a minute! I know that if the ant world is being inflated, then the ants should notice! After all, I notice when, say, my mom hits the gas in the car." True enough, but in this case the ants wouldn't notice because their universe is expanding in a mysterious third dimension of which they aren't directly aware.†

Perhaps we're moving in a fourth spatial dimension, which would be different from the three spatial dimensions we're accustomed to. Later

---

*They would also probably note, "Holy ____, I'm burning alive!"

†Anyway, ants can't drive.

in this chapter we'll discuss the possibility that there might be other dimensions beyond the three we're directly aware of . Then again, this also might be a case of trying to carry an analogy too far. In our current Standard Model of cosmology, we don't need anything beyond three spatial dimensions (plus one for time).

What's at the edge of the universe?

Our discussion of Tentaculus VII brings up an important point. Provided we had telescopes powerful enough to see Dr. Snuggles's home planet, the picture we would see is not how it is today, but how it was about 1 billion years ago. We might look at another, even more distant galaxy and note that we're looking at it even farther in the past.

Astronomers are able to study the properties of galaxies in the very early universe by looking at galaxies that are very far away.

But as we look at more and more distant objects, there is a distance beyond which we can't see. On Earth we'd call this the horizon, and it's no different in the universe as a whole. We can't see beyond the horizon because light moves along at a constant speed. Since the universe has been around for only a short while—about 13.7 billion years—anything farther than about 13.7 billion light-years away won't be visible to us for some time.

Where'd this so-called beginning of the universe come from? We'll do our reasoning backward. If all the galaxies in the universe are currently moving away from one another, there must have been some time in the past when they (or at least their constituent atoms), must have been right on top of one another. We refer to this event as the Big Bang, and it's the subject of a lot of confusion (and the next chapter).

We can estimate when the Big Bang occurred by remembering that speed is just the ratio of distance over time. Assuming (wrongly, as it turns out, but close enough for now) that the recession speed of the Tentaculus home galaxy hasn't changed since the beginning of time, then the age of the universe can be computed with a simple bit of mathemagic. Think about it: the farther away a galaxy is today, the older our universe must be, since everything is moving away from everything else at a measured rate. Plugging in the numbers for our universe, a back of the envelope estimate for the age of the universe is about 13.8 billion years, very nearly what you get if you do the calculation correctly.

If we had a powerful enough telescope, could we see the beginning of the universe? Almost, but not quite. The current distance record-holder, a stellar explosion dubbed GRB 090423, is so far away that the image we see in the Swift satellite is from when the universe was only about 630 million years old (about 5% of the current age of the universe), at which point the universe was less than a ninth of its current size.

Stranger still, GRB 090423 appears to be moving away from us at approximately eight times the speed of light. (We'll wait while you flip back to chapter 1, in which we clearly state that this is impossible.)

This mystery goes away if we remember that it's the universe expanding and not the star moving away from us. The star is essentially standing still.

Seem like we're cheating? We're not. Special relativity doesn't say that things can't move apart faster than the speed of light. What it *does* say is this: if I shine my Bat-Signal into the sky, Batman can't outrace it in his Batplane, no matter how hard he tries. More generally, it means that no information (a particle or signal, for example) can ever travel faster than light. This remains absolutely true, even in a quickly expanding universe. There's no way we can use the universe's expansion to outrace a beam of light.

We actually can see even farther back in time than GRB 090423, but we need radio receivers to do so. We can see back to when the universe was only 380,000 years old and consisted of nothing more than a swirling amalgam of hydrogen, helium, and very-high-energy radiation.

Beyond that distance, things quite literally get hazy. Since the universe was thick with stuff in its early days, it's like trying to see through your neighbor's curtains.* We can't see what's on the other side, but we do know what the universe looks like now (and all the time between the early universe and now), so we can guess what's behind that cosmic curtain. Tantalizing, isn't it?

So while we can't quite see back to the horizon, we can get close enough for government work. The really cool thing is that the longer we wait, the older the universe gets and the farther away the horizon gets. In other words, there are distant parts of the universe whose light is only now reaching us for the first time.

What lies outside of the horizon? Nobody knows, but we can make an educated guess. Remember that Copernicus and his successors have shown us effectively "Wherever you go, there you are," so we might suspect that the universe looks more or less the same outside the horizon as it does here. Sure, there will be different galaxies, but about the same number as there are locally, and they'll look about the same as

---

*Not that you would ever try to do that. We don't like to think of you that way.

our neighbors. This doesn't absolutely have to be true. We assume it because we have no reason to believe otherwise.

 ## What is empty space made of?

So the universe is expanding, but the actual galaxies in the universe are barely moving. We need to return to Einstein's theory of general relativity. John Archibald Wheeler famously described the theory by saying, "Space tells matter how to move, and matter tells space how to curve," and this is exactly how you should think of it.

General Relativity
"The Universe is highly classified."

We remind you of our promise to steer clear of math, but Wheeler's description is actually a succinct way of describing the central equation of general relativity, the Einstein field equation. While we won't write down the field equation, we will say a few things about it.

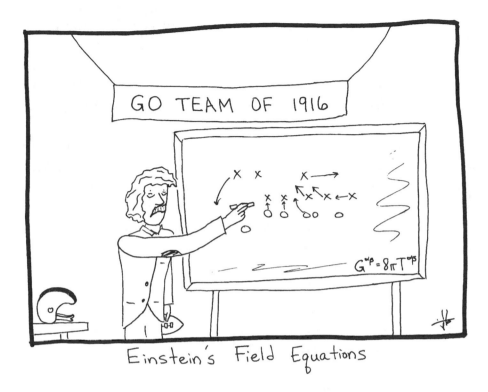

Einstein's Field Equations

The left side* of the field equation determines how far apart two points are from each other in both space and time, a quantity known as the metric, and by exploring how the metric changes over space, we're able to describe how curved space is. The metric is important because the particles are lazy, and will take whatever route minimizes their travel time. In flat space (that is, without gravity), the quickest route is a straight line, as you would probably guess, but if space is curved due to gravity, things get more complicated.

Let's say you throw a ball to a friend. The ball wants to get to your friend as quickly as possible, so maybe the quickest path is a straight line. But wait! Gravity (as we saw in the previous chapter) makes time run a *tiny* bit slower near the surface of Earth, so the ball might get to your friend quicker if it gets away from Earth for a bit and travels in an arc.

---

*Yes. We are, in fact, writing down things such as "the left side of the field equation" as a refuge from writing down the equations themselves. Don't get all bent out of shape.

On the other hand, if it arcs too much, it will have to travel faster, and as we saw, time also passes slowly for a fast-moving ball. It compromises, and by following the curve of space-time, the ball appears to travel in an arc. See? Despite all of this talk about relativistic time and bended space, in weak gravitational fields such as Earth's, gravity behaves just like Newton said it would.

But we're going to have to escape Earth's weak fields if we want to figure out how the universe evolves as a whole, and to do that we need to say a thing or two about the metric. Remember that the metric is what tells us how far apart two points are from each other. Imagine you had a ruler that was slowly shrinking. As you made subsequent measurements of the distance between, say, yourself and Poughkeepsie, you'd find that the distance was constantly increasing.

This is precisely what's happening in the real universe!

Space is not the absolute thing that we were taught in grade school. We already saw that space and time are relative for moving observers and for observers near massive bodies. We now realize that space itself changes as the universe ages.

And what of the right side of the Einstein field equation? Wheeler already gave us the answer: "Matter tells space how to curve." The matter in the universe tells the universe how to evolve.

How will we be able to understand all this when we don't really know the equations of general relativity? Have no fear. Remember that your physical intuition about gravity works better than you'd expect.

We've rather glibly been talking about the expansion of space without saying anything about what space really is. Isaac Newton thought a lot about space in his *Principia Mathematica*, and devised a little thought experiment to make things concrete. Let's think way back to chapter 1, in which Rusty, Galileo, and Einstein (though not necessarily in that order) found that no observer can tell if he or she is moving or at rest, as long as he or she moves at a constant speed. The only important thing in the dynamics of two observers is their relative state of motion.

Newton imagined a bucket hanging from a twisted rope, filled with water, and held still. The bucket is then released, and starts to spin while the rope untwists. At first the water wants to sit still, and the sides of the bucket spin around it. Eventually the friction between the bucket and the water kicks in, and the water spins with the bucket. As it does so, the water creeps up the sides.

We know you're thinking, "So what?"

The reason why this is kind of a big deal is that by the end of Newton's bucket experiment, there is no relative motion between bucket and water, and yet we can still tell that the bucket and water are twisting. The real question is: how does the bucket "know" that it's spinning?

As an example of something you can see if you visit just about any science museum, consider Foucault's pendulum. A pendulum is just a weight (or bob), suspended by a rope or a rod, and allowed to rock back and forth, like inside traditional grandfather clocks. In the case of Foucault's pendulum, the device is mounted so it is free to move in any direction it wishes. The bob is set to rock back and forth, but an observer watching long enough will note that it also rotates. Or rather, the pendulum just moves back and forth while the Earth rotates underneath it. Somehow it knows to maintain its fixed orientation with respect to space.

Or consider our old pal Rusty in deep space, sitting in a big rocket-powered chamber, much like the gravitron at the amusement park.

The rockets start up, and the cylinder starts spinning. After a short while they stop, and the whole contraption just keeps on spinning. If you've seen *2001: A Space Odyssey* or any other sci-fi movie in which artificial gravity is simulated in a spinning space station, you know that Rusty will be thrown up against the walls.*

If it's just Rusty and his gravitron, alone in the universe, we have a bit of a quandary. How can they be said to be spinning at all? Spinning with respect to what? Try to answer the question without using the word "space." Space is just nothingness, after all.

The philosopher Ernst Mach addressed this same question in his *Science of Mechanics* about 240 years after Newton did:

> [The] investigator must feel the need of . . . knowledge of the immediate connections, say, of the masses of the universe. There will hover before him as an ideal insight into the principles of the whole matter, from which accelerated and inertial motions will result in the same way.

This explanation isn't what we'd call a precise scientific statement about how the universe works, and it seems reasonably likely that we would have forgotten what Mach had to say about the matter if it weren't for the fact that Einstein was obsessed with "Mach's principle" (as Einstein coined the idea). He rephrased it much more succinctly: "Inertia originates in a kind of interaction between bodies."

Still too complicated? How about, "Mass *there* influences inertia *here*"?

Well, duh! Of course matter far away affects the motions of bodies here. That's just what we call gravity. But that's not what Mach was saying, nor what Einstein read into it. What Mach was saying was that somehow by comparing our matter to the distant stars, we can figure out whether we're moving, or at least whether we're accelerating. It's not that different from you saying that your train is moving by observing the apparent motion of mountains. The mountains are big and you are small, so your motion can be measured relative to the big stuff in the universe.

---

*Or if he stays on the ride too long, he will be *throwing up* against the walls.

Einstein used Mach's principle as one of the chief inspirations for his theory of general relativity. His basic idea was that the "distant stars" could be fixed on average, and that you could only really say that something is accelerating—or rotating, for that matter—if it was doing so with respect to the fixed stars.

Is Mach's principle right?

It doesn't have to be. Mathematically, there is a solution to Einstein's equations for empty space: that is, no matter at all. Clearly there are no distant stars in such a case, but Einstein's theory of special relativity would still predict that if you suddenly popped into that otherwise empty universe, you could "feel" if you were rotating.

But a completely empty universe clearly is the exception rather than the rule. Our universe has stuff in it. General relativity incorporates the matter in the universe. It's this "warping" of space that can be felt throughout, including here.

Almost immediately after Einstein came up with his theory of general relativity, Josef Lense and Hans Thirring of the University of Vienna noted that if you take something massive enough—say, a black hole—and spin it, the space around the black hole will get dragged around as well. To put it another way, if you tried to stand still, it would seem as if you were spinning. This idea isn't just a guess, either. A number of satellites have since been launched that have measured the dragging of space due to Earth's and Mars's spin.

The point is that on the biggest scales, it seems that matter is what "makes" space, even if the local space looks like there's nothing in it.

 ## How empty is space?

We've been veering a bit too much toward the esoteric in the past few pages, focusing on the nature of space and the like. Now it's time to get a little bit more concrete. To that end, we'll make you a deal: if you agree that the galaxies in the universe are basically sitting still while the universe expands around them, we'll admit that there's often no real harm indulging in the fantasy that we're at the center of the universe. If you agree, please signify by shaking this book vigorously.

We'll take that as a "yes."

We can even do a fair bit of accurate physics using the you-centered model. Let's start with the basic question of whether the expansion will slow down or speed up. Look at it from the universe's perspective and try the following experiment:

1. Go outside with a baseball.
2. Throw it straight up into the air.
3. Get out of the way.

No matter how many times you repeat the experiment, eventually the old adage comes true—what goes up must come down.

Of course, the reason we're able to build rockets that can go to Mars is this: if you throw the ball or boost the rocket fast enough, it can escape Earth's gravitational pull. The escape velocity of Earth is about 25,000 mph. Rockets get into space because they can move faster than that.

On the Moon, however, the escape velocity is a bit over 5,000 mph. In fact, if you were on the Moon and you could throw a 10,000 mph fastball, you'd find that the ball would go clear to outer space. A throw of the same speed on Earth would result in the baseball crashing back to the ground. To put this in even more perspective, Mars's moon Deimos has an escape velocity of about 13 mph. Even *we* could throw a ball with escape velocity from there! Probably.

What makes Deimos so different from Earth? Mass.* Earth has lots more of it, and thus more gravity. With less mass, there is less gravity pulling the baseball back to the planet (or planetoid, or moon, or whatever), and consequently Deimos has a smaller escape velocity. This maxim holds for massive objects such as galaxies as well.

If the universe were completely empty (it's not, thankfully for us), it would continue to expand forever, its rate completely unabated. There would be no matter to slow it down. If we had such an empty universe and we put a little bit of stuff in it, the expansion would slow a little bit. Remember that matter affects space, so if we put a whole bunch of stuff in, the universe would eventually collapse back on itself.

---

*That's right. Deimos is Roman Catholic.

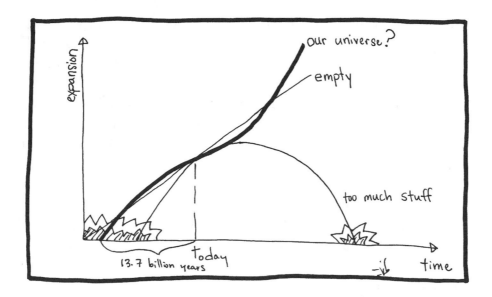

The dividing line between the fate of infinite expansion and that of a big crunch is called the "critical density" of the universe, and it's much lower than you might think.

People tend to have an overly inflated sense of how packed outer space is, so perhaps a reality check is in order, and we'll start locally. Think back to the scene from *Star Wars* in which Han Solo is barely able to hold the Millennium Falcon together as he makes his way through a crowded asteroid belt. As you probably know, our own solar system has an asteroid belt between the orbits of Mars and Jupiter (the fourth and fifth planets from the Sun, respectively). What would happen if you were so foolhardy as to try to pilot your ship to Jupiter?

Not very much.

While astronomers aren't exactly sure how many asteroids there are, a conservative estimate of ten million or so means that the average distance between these guys is more than a million miles apart. If you don't have a feel for that sort of distance, a million miles is about four times the distance to the Moon, a trip only a couple of dozen human beings have ever undertaken.

As we head out of our solar system to other stars, more than four light-years separate us from the nearest star, Proxima Centauri, and between here and there, things are really quite barren. On average, every cubic centimeter (about as big as a regulation-size die) of interstellar space contains only about one hydrogen atom. That's about 10^{16} times emptier than air here on Earth, and about a million times less dense than the best artificial vacuums we're able to build.

Between galaxies, even if the universe is at the critical density, space is another factor of a million less dense. That works out to only about five atoms of hydrogen for every cubic meter (roughly the size of your refrigerator) of space.

Of course, you probably expected that outer space is empty. That's why it's called, you know, space.

Because astrophysicists don't like dealing in the small change of atoms, we really only care about whether the universe is more or less than the critical value, so we define a ratio. This ratio compares the amount of matter (any kind of matter) in the universe to the amount of matter we would expect at the critical density. We call this ratio

$$\Omega_M.$$

If you want to tell your mom about what you learned in this book[*] and don't want to write anything down, this is referred to as "omega matter."

We're going to ruin the surprise and tell you that our best estimate of Ω_M is 28% (plus or minus some smallish error) of the amount of matter necessary to make the universe collapse on itself. As the universe expands, the matter in the universe gets more and more diffuse, and so as time goes on it will start to look more and more empty. This means that the density of the universe will decrease (space is getting bigger, but no more matter is being produced), so this ratio will get smaller and smaller.

It's a pretty important number (at least among geeky astronomers), and over the past twenty years or so, the bulk of mainstream cosmology has focused on trying to get this number and just a handful of others[†] to figure out the age, fate, future, and past of the universe. This number in particular is important, since it tells us whether the universe is going to recollapse, or continue expanding forever. To determine its value, we

---

[*]"Mom? Yeah, it's me. I'm reading this book about physics, and I wanted to tell you how dense you are compared to the rest of the universe."

[†]This is *by far* the most successful attempt that astronomers have made in getting numbers.

need to measure how much stuff is around us, and so the basic question is, How do we weigh the universe?

There are about a hundred billion galaxies in the observable universe, and they contain most of the mass. If we can figure out how to weigh galaxies, or clusters of galaxies, then we can just add up all the mass within a particular region of space, and then we will know the density of the universe.

Where's all of the stuff?

Instead of trying to weigh the entire universe, if we figure out a way to effectively weigh individual galaxies, we're golden. So how's this for an idea? Count up all the stars in a galaxy and assume that they're pretty much all like the Sun. After all, when you look out into the night sky, everything you see is just starlight, or, in the case of the Moon and the planets, reflected starlight from our own Sun. What's more, in our own solar system, about 99.99% of the mass is in the form of stars (our Sun), so maybe assuming that (virtually) all of the mass in galaxies is in the form of stars isn't so crazy. If we crunch the numbers in our fancy computing machine we find that the universal density, Ω_{STARS}, is only about 0.2%.

This result means that, like the Autobots and Decepticons, there must be more to galaxies than meets the eye. The dominant contribution of ordinary "stuff" in most galaxies is huge amounts of gas, which radiates in X-rays rather than in visible light. So if you could somehow take your favorite galaxy to your dentist's office, he or she would be able to tell you how much gas is in it by measuring the X-ray radiation. If you include mass measured through this effect and add it to all of the mass from stars, you'd find that Ω_M is about 5%, which still suggests a universe that is pretty darn empty.

That 5% is kind of surprising, and more than a bit troubling. It represents the amount of mass contained in ordinary stuff, what physicists like to call baryons, which you remember[*] are just protons and neutrons. This means that all of the elements are made of baryons, which means that all atoms and molecules are made of baryons, which

---

[*]Sure you do.

means that you and me, the Sun, Earth, stars, gas, dust, and everything you've ever seen or had contact with are made of baryons. There are a bunch of different tests you can do to count up the baryons in the universe. All of them suggest that Ω_B, the fraction of the critical density in baryons, is only about 5%.

That would be all well and good except for a curious observation first noted by Vera Rubin and her collaborators in 1970. She noted that stars are orbiting around galaxies, and the whole thing is held together by gravity. If a galaxy doesn't have enough mass, then the stars would fly off. This effect is exactly what happens if you do the classic yo-yo trick "around the world" and someone cuts the string. The yo-yo no longer stays "in orbit" but rather flies off, presumably knocking someone's teeth out.* The point is that we can measure how fast stars are moving around the centers of galaxies by calculating their Doppler shift, and from that we can measure the total masses of the host galaxy. And you know what? The galaxies are about six times more massive than we would have guessed! In other words, Ω_M is about 28%, but only if we assume that most of the mass—about 85% of the total—is made up of some sort of mysterious "dark matter" that we can't see.

Maybe there is just something wrong with our measurements, or maybe we're doing the calculation wrong. Occam's razor tells us that we should take the simplest solution as the best one, and it's much simpler to say that we're making some sort of mistake than to say that somehow we're just not able to see 85% of the mass in the universe! We need other tests.

In recent years, good-looking, intelligent astronomers have started using a technique known as gravitational lensing to measure the masses of galaxies and clusters of galaxies. Lensing exploits the fact that massive things such as galaxies bend space a bit, and light beams follow the bending of space. For example, if the Tentaculus home galaxy lay between Earth and a more distant galaxy, then the image of the background galaxy will be distorted by the mass of the Tentaculan galaxy. The greater the mass, the more the distortion.

---

*Between your dangerous yo-yo stunts and galaxy X-rays, this chapter is going to cost you a fortune.

For clusters of galaxies this effect can be quite dramatic, since clusters can have masses as large as a quadrillion (10^{15}) times the mass of the Sun. When clusters lens background galaxies, these otherwise normal-looking background galaxy images can be seen on Earth as grotesque arcs, and occasionally there will be two images of the same galaxy in much the same way that a magnifying glass can be held so that more than one image of, say, your finger can be seen.

As one of the rare cases when a cartoon just isn't going to cut it, take a look at a Hubble Space Telescope image of the cluster Abell 2218:

If you look, you'll notice some very bright, roundish galaxies. Those are the galaxies in the cluster. However, you also may note that there are a bunch of really elongated smudges and dramatic arcs. Believe it or not, those are also just ordinary galaxies, but they're behind the cluster (as seen from Earth), and their images are grossly distorted by the gravitational field.

Lensing provides another way of measuring the masses of galaxies, and, consequently, the universe, and the numbers point to the same thing: there's about six times more mass in the universe than can be accounted for by "ordinary" baryonic mass. In a particularly stunning result, in 2004, Douglas Clowe, then at the University of Arizona, and his collaborators studied a colliding pair of clusters known as the "bullet cluster" and discovered something remarkable.

As we just saw, the majority of ordinary mass in clusters is not made up of stars; it's made of hot gas. The stars, the part of the galaxies that we can see with our eyes, are just a small minority. So if this hard-to-see matter, or dark matter, were really made of ordinary stuff, we might expect it to line up with the gas.

What Clowe and his collaborators found was that not only is more mass in the cluster than one might guess based on the gas, but also the dark matter didn't even appear to be near the gas! In other words, even though we don't know what it is exactly, we now know how to find it. We'll return to the question of what exactly dark matter is in chapter 9.

 ## Why is the universe accelerating?

The quest for dark matter almost completely defined the state of cosmology until about 1998. Because the results of measuring the masses of galaxies were still coming in, much of the cosmological community was convinced that Ω_M had to add up to 100%. There was no strong evidence to the contrary, and most theories favored that number.* However, a series of observations in the mid-1990s blew that idea apart.

A bit earlier, we mentioned the idea that one of the chief ways that we're able to measure the distance to other galaxies is that we know how bright they are intrinsically, and by measuring how bright they appear to us, we're able to measure their distances. Nature seems to have provided us with an excellent "standard candle" in the form of a type of exploding stars called "Type Ia supernovas."

Type Ia's are comprised of a white dwarf and a red giant star, which are in orbit around each other. The white dwarf is a smoldering core

---

*The theory was and is known as inflation, and we'll talk a lot about it in the next chapter. Current estimates show that Ω_M is about 28% (as opposed to the 100% predicted by the first models of inflation), which should have demolished inflation once and for all. Somehow, though, the theorists realized that they could tweak their equations to make everything work out. This should teach you a valuable lesson about believing a theoretical physicist who claims to be certain of something.

of a star, and quite dense. The red giant star is big, and its gravity is relatively weak, which means that the gas from the outer atmosphere of the red giant falls onto the surface of the white dwarf.

White dwarfs are very compact objects. When our Sun turns into a white dwarf, it will become as small as Earth.[*] These stars are so dense that their individual electrons are literally bumping against one another. The densities in white dwarfs are about a million times that of rock, and it is very, very hard to compress a white dwarf any more. However, eventually enough of the red giant's castoffs fall onto the surface of the white dwarf that it can take no more, and the protons and electrons in the star merge to form a neutron star. When this happens, there is an enormous shock and explosion of

A RED GIANT ACCRETING ONTO A WHITE DWARF

---

[*]Of course, we on Earth, having long since been fried to a crisp by the Sun having previously become a red giant, will be in no position to care.

material called a type Ia supernova. During a few weeks, as much energy is given off by this explosion as the Sun will emit in its entire ten-billion-year lifetime.

You don't want to be near a supernova when it goes off. Even if one went off ten light-years away, the results would be fatal for life here on Earth. Fortunately, any given galaxy has only about one of these explosions every century, and our own Galaxy is tens of thousands of light-years across, so odds are that we're safe for the time being. The bad news, though, is that there's no way for us to predict when or where a supernova will go off.

But astronomers (typically misanthropes) still love these astrophysical cataclysms. Supernovas make superb standard candles because (1) they are incredibly bright, which means that they can be seen over huge distances, and (2) because they all "go off" at about the same time (that is, when a certain amount of material has fallen onto the white dwarf), and they all look more or less the same, which means that we can calibrate how far away they are.

In 1998, two teams, led by Saul Perlmutter and Adam Reiss, respectively, measured distances to approximately fifty of these supernovas, and since they also could collect redshifts, they not only knew how far away the supernovas were, but also how much the universe has expanded since then.

They each, simultaneously and independently, found something remarkable. The universe isn't slowing down, as one might expect from pretty much everything we've told you so far. It's accelerating. Einstein had stumbled on something of this sort when he originally came up with the idea of general relativity. Einstein called it the "cosmological constant," and if you've ever done calculus, it's very much like the "plus a constant" that comes out of integrals. If you haven't done calculus, you'll have to take our word for it.

Einstein invented the cosmological constant as a way of making the universe static, and he was greatly embarrassed when Hubble discovered the expanding universe. However, despite its origins, the math behind the cosmological constant is sound, and after the supernova results came out, there was renewed interest in a cosmological constant. This time around, though, the constant was thought of as a "dark energy" that pervades the universe.

Einstein noted that gas with high pressure has a stronger gravitational pull than gas with no pressure at all. This difference is important because dark energy has negative pressure; it behaves like a sort of antigravity and causes the universe to accelerate. Even stranger, as the universe expands, the density of the stuff doesn't decrease. It's exactly as if you had a ball of taffy that you stretched and stretched but somehow it didn't get any thinner. This instance is definitely one of those cases where your physical intuition is likely to let you down.

Think this sounds too weird to be true? Well, it isn't. We've already seen something very similar in chapter 2. Remember that the universe is pervaded by a "vacuum energy" due to the fact that photons keep popping into and out of existence. Remember, also, that if we stretch or smash a box of this vacuum energy, the density stays the same.

We know, it seems like we're just playing games, so perhaps you'll be reassured that such an effect is actually observed. In 1948, Henk Casimir of the University of Leiden noted that if you take two metal plates in a vacuum and hold them a small distance from each other, the two will, surprisingly, attract each other. That shouldn't happen if the two plates aren't electrically charged. It all makes sense if we assume that there's a vacuum field everywhere in the universe. Since electrical fields vanish on the plates, the vacuum field will be lower between the plates than outside, and as a result the plates will be pushed toward each other.

The "Casimir effect" is one of the strongest and most direct pieces of evidence that the vacuum energy is a real thing, and has exactly the properties we're looking for in terms of dark energy.

That's the good news.

We get the bad news when we ask how much dark energy there seems to be in the universe. Since matter and energy are equivalent (as we saw in chapter 1), we can ask what the density parameter of dark energy is, and we find that Ω_{DE} is about 72% from cosmological measurements. We use the subscript "DE" to remind you that we're talking about dark energy.

The Kashmir Effect

This number seems like more good news, because if you add

- $\Omega_B = 5\%$ (ordinary stuff)
- $\Omega_{DM} = 23\%$ (dark matter)
- $\Omega_{DE} = 72\%$ (dark energy)

we find that the total energy density in the universe is the critical value—Ω_{TOT} (what you get if you just add up the different contributions), 100%. This result is going to have some pretty interesting implications.

Now for the bad news. If we understand the experiment with the metal plates correctly, then both the laboratory experiments and most theories suggest that the vacuum energy in the universe should be about 10^{100} times larger than we see from cosmological measurements.

In physics, we call that a problem.

 ## What is the shape of the universe?

The reason we are making such a big deal about Ω_{TOT} is that the density of the universe does more than just tell us how the universe will evolve; it also describes the shape of the universe.

Here's what we mean. Both Earth and Tentaculus VII are, as we have said, more or less fixed in the universe. Far away, a billion light-years from each, there is a civilization of hyperintelligent robots, the Klankons, led by their leader, the Astronomer King, XP-4. By an astonishing coincidence, on one particular day, Hubble, XP-4, and Dr. Snuggles each take an image of the other two star systems, and record the angle between the two.

Wait! Where do angles come into it? When you look into the night sky, you might realize that you don't have a true 3-D picture of the universe. Stars that are near each other in the sky may be physically near one another, or their proximity may be a coincidence, with one far away, and one close. On Earth, we can resolve these ambiguities through the magic of our binocular vision (depth perception that we get from having two eyes), but for distant galaxies, we simply can't tell, so the only measurement we get just by looking is how far apart two stars or galaxies are in angle.

Now, continuing the convoluted experiment, all three civilizations transmit their angular measurements to the other two. Each of them now (or a billion-odd years from now) knows all of the interior angles of an equilateral triangle in space.

If we drew such a triangle on a sheet of paper, we know that each angle would be sixty degrees. This is the epitome of what would happen in flat space and, as it happens, space will be exactly flat if Ω_{TOT} equals exactly 100%. Flat universes are comforting to live in because your intuition seems to hold up rather nicely.

But flat universes are not the only possibility. Remember what Wheeler told us about matter telling space how to curve? If Ω_{TOT} is greater than 100% (as it might have been if there were lots more "stuff" in the universe), then cosmologists say that the universe is "closed." It's pretty easy to imagine a closed geometry. It behaves almost exactly like the surface of Earth. Connecting three points with a triangle, we'd find that the interior angles add up to more than 180 degrees.

We apologize for our rather poindexterish display of geometry, but we have to point out one more cool thing about our triangles. Take a galaxy, place it far away from Earth in a flat universe, and then (provided parallel universes exist) do the same thing in a closed universe. The galaxy will appear bigger in the closed universe.

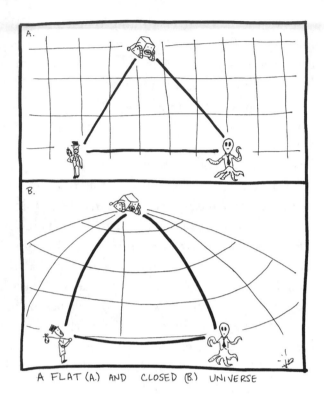

A FLAT (A.) AND CLOSED (B.) UNIVERSE

It's time for a quick quiz. If universes with Ω_{TOT} bigger than 100% are closed, what do you suppose those smaller than 100% are? If you said "open," we'll be happy to send out your Ph.D. in tomorrow's mail. As expected, open universes have the property that distant objects appear smaller than you'd expect from a flat universe.

Which one do we live in? If our cosmological parameters are to be believed, then we live in a flat universe, or at least something very, very close to it. In reality there's not much of a practical difference between nearly flat and flat. It's kind of like being on the surface of Earth. Yes, Earth is round, but in your day-to-day activities you can easily forget that fact.

A closed universe is the only one of the three that is finite. That's not to say that you can walk to the end of it. Just like a sphere, if you walk around forever and ever you'll eventually come back to where you started, but you won't ever hit the edge.

We normally think of flat (and open) universes, on the other hand, as infinite. It's a little tricky to say what exactly infinite means, but it's certainly the case that the universe has no edge. It also may mean that the universe is quite literally infinite—that is, you could keep traveling forever but never hit the same place twice.

Or maybe not.

What general relativity describes is the so-called geometry of the universe. A piece of paper, if folded up into a tube, is still a piece of paper, which means that it's still "flat" from a geometric point of view. Everything we talked about with triangles earlier works on a rolled-up piece of paper.

Like a flat piece of paper folded up into a tube, it may be that the universe is rolled up back on itself. This concept illustrates what is known as the topology of the universe, but we have no physical theory that tells us if or how the universe rolls back on itself.

In principle, Dr. Snuggles could look into the night sky and see two images of the Klankon home star on opposite sides of the sky. In 1998, Neil Cornish of the University of Montana and his collaborators looked to see if there was a similar effect measurable in the signal from the microwave background, the remnant of the Big Bang. No such signal was found. This outcome doesn't mean that the universe doesn't fold on itself, but if it does, it does so on scales much, much larger than the horizon.

What's the universe expanding into?

All of this talk about dynamics and geometry might seem a bit beside the point. But we're now prepared to figure out what the universe actually expands into. The problem is that general relativity and our observations can't really answer the question. Remember, physics only tells us what happens under certain circumstances, not about what the universe is really like at the fundamental level. And cosmology presents its own special problems. We have only one universe to observe. If you don't like the answers, maybe you're not asking the right questions.

So while it may seem a bit disappointing, we're afraid we can't give you a definitive answer. We can, however, give you a bunch of different ways of thinking about the issue.

What's the universe expanding into? Take your pick.

1. Nothing

In our opinion, this is the best answer. If we think back to how general relativity works, the metric—how far apart any two points are from each other—is the only thing that defines how space works. As a result, there is no "outside" of the universe. This is the perspective that we've given through this entire chapter. You could keep on flying and flying, and you'd never hit the edge. Even if the universe were finite, the whole thing just might fold back on itself.

2. It Doesn't Matter

We realize that this is a nonanswer. The point is that the only physics we can ever observe is what's going on inside our horizon. It is conceivable that outside the observable universe is nothing but an empty void, completely devoid of matter. Perhaps everything is purple, or there are other "island universes" with properties very different from our own. We have no idea. If it's outside our horizon, we can't ever know. Remember that Copernicus's assumption of nonspecialness states that the universe is the same no matter where you are, so the odds are that we aren't missing anything special.

On the other hand, as the universe continues to expand, our horizon gets larger and larger, we'll see more and more, and we'll get a better and better glimpse into whether we are or aren't in a special place in the universe. Within limits.

As it happens, our universe has dark energy in it and, as time goes on, the ordinary matter and dark matter get more and more diffuse, but the dark energy doesn't. They will just keep accelerating and accelerating, which means that any given point in space will keep rushing away from us faster and faster. In our universe, the horizon eventually will reach some maximally distant horizon. If there's anything outside of that, we can't ever know about it.

3. Higher Dimensions?

We mentioned the possibility that there are potentially other dimensions in the universe besides the up-down, left-right, and forward-backward that we are accustomed to. Time is certainly another dimension, and in a sense, the universe is expanding into time, but as a physical theory this explanation is mumbo jumbo. The universe is moving into the future in the same sense as everything else.

The past few decades have seen an explosion of models of the universe with more than three dimensions, more commonly known as string theory, culminating with the ten-dimensional M theory, which we last encountered in chapter 4. As you'll recall, according to string theory, the differences between or among particles are all in your head. At their heart, all particles are basically strings, and one string can split off into two,* or two could be tied together into one.

But M theory, in particular, also predicts some more complicated structures. While the "string" in string theory really only accounted for one-dimensional structures, the strings themselves, M theory predicts the existence of much more complicated, two- and three-dimensional "branes" (short for membranes). Individual particles (such as photons) could be "stuck" to a particular brane.

The connection for us is that it's possible that our entire universe is simply a gigantic three-dimensional brane and that we're moving around in higher-dimensional space. Perhaps there are other "universes" hovering nearby, but since our photons are trapped on our own brane, and their photons are trapped on their brane, we never see them. M theory suggests, however, that we could feel them or at least their gravitational effects, and perhaps these branes occasionally collide, causing an end and rebirth of our "universe."

---

*You could do this with a regular loop of string by cutting it in two, then tying off the two pieces into new loops.

ZOMBIE STRING THEORISTS

In that sense, perhaps our universe, our brane, might be moving into the higher-dimensional universe.

So at the end of the day, the universe seems to be expanding into nothing. Of course, it may yet turn out that the "outside" into which we're expanding (or at least moving) are higher dimensions that we can't possibly experience directly. How's that for a "blow your mind" picture of the universe?

The Big Bang

"What happened before the Big Bang?"

☒ PICTURE NOT AVAILABLE

The Big Bang   jb

We at the *User's Guide* are not yet parents, but we've heard stories. One of the most uncomfortable conversations you can have with a small child (or so we're told) begins with Little Billy asking, "Where did I come from?" When and if this day comes, we have a plan already lined up. We're going to bide our time by starting at the *very* beginning,[*] and we will talk about pirates, because kids *love* pirates.

We also like to flatter ourselves that our kids are going to be the sort who can understand complicated discussions about the expanding universe, Grand Unified Theories, and the origin of matter while they're still knee-high to a grasshopper. No baby talk for us; just the creation of the universe and adventures on the high seas!

While we could start at the beginning and arrive at the here and now, it makes more sense to explain the universe in the reverse direction. This story starts at the end, and documents how everything came to pass.[†]

---

[*]Admittedly, approaching the subject with words like "the Big Bang" could have a detrimental effect later, when Little Billy is old enough to know where he *actually* came from. We leave this matter to the psychologists.

[†]If this were *Jeopardy*, the question would be "What is 'a diet with plenty of fiber'?"

Our brave pirate captain, Bloodbeard, has just been defeated by the Spanish Armada and has heroically gone down with the ship. A few of his men were not so brave and escaped in boats, setting out in all directions. Some rowed away quickly; others, slowly.

A latecomer to the scene might see only the escape boats (Bloodbeard long since having been sent to Davy Jones's locker), but if he or she were clever enough, the latecomer could figure out that all of them must have originated from the same place. By noticing how far away the cowardly shipmates were able to row, the onlooker would even be able to tell how long ago the battle took place.

This story is a metaphor, you see. The boats are really supposed to represent galaxies, and as we saw in the previous chapter, almost all galaxies are flying away from one another. It's reasonable to suppose that once upon a time, all of the galaxies were literally on top of one another, just like lifeboats on a pirate ship.

Like most tall tales, while our pirate story has a kernel of truth, it's also got a glaring error. While it'd be easy to tell you that the galaxies are sailing away from a common point in space, we can't. What we can say is this: the Big Bang happened everywhere, all at once. This is important, since almost everyone (not just Little Billy) assumes that the Big Bang must have happened somewhere in particular. We saw the same thing in chapter 6 with the expansion of the universe—space gets larger and larger without any of the galaxies actually moving away from one another.

There's a second detail that we're glossing over in our story. The universe wasn't created with the galaxies already prebuilt. Instead, we start with just gas and dark matter. It's as if the cowardly sailors abandoned ship with kits from IKEA and slowly constructed their boats as they floated away. The main tool in the galaxy-building toolkit* is gravity. You'll remember from chapter 4, if not from grade school, that all matter attracts all other matter. Shortly after the Big Bang, some regions of space had a little more stuff than others, and if we watched a small lump that was a tiny bit denser than average, we'd see something interesting. The nearby gas and dark matter would get attracted, and slowly that little lump would grow into a big one, ultimately forming into the galaxies we see today.

But the basic point remains: those atoms (and dark matter and dark energy) that make up everything we can (or can't) see were once *basically* on top of one another, and now we have to explain what happened between then and now. At the beginning, we start with a universe that was infinitesimally small. You can explain that this is where Little Billy came from: the Big Bang.[†]

Little Billy is precocious, and will point out that we didn't really answer the question at all. If the Big Bang was the cause of the universe, what caused *that*? Let's one-up Little Billy with an even more fundamental question: how can we even be so sure that there was a Big Bang in the first place? It's not like there was anybody there to observe it.

---

*The equivalent of the little bendy hex wrench.
[†]Do you understand now how this could scare or even scar a child for years? Please, use science carefully.

What's more, even though we're able to look into the past by looking at more and more distant objects, we can't actually *see* the Big Bang, so any evidence we might have is purely circumstantial. This is one of the main reasons why we'll start with what we know.

According to our best estimates based on the expanding universe theory, the universe is about 13.8 billion years old, and right now, space, as we argued in the previous chapter, is largely empty. There is, however, a heck of a lot of space, and plenty of *stuff* spread throughout—just very diffusely. Besides dark matter, dark energy, stars, dust, and gas with which you are already familiar, the universe is jam-packed with light. Oh, sure, it looks dark out there, and you're probably fooled into thinking that all of this light originates from the obviously bright and shiny things such as the Sun. Don't be. The contribution of starlight (including that of the Sun) to the total number of photons in the universe is very small. For every atom in the universe, there are about a billion photons, and these photons—or the vast majority of them—have been around since nearly the beginning of time.

Despite the huge number of photons flying around, most of the time we aren't normally aware of them, since while there is a lot of background radiation, it is at extremely low energies. This is a consequence of the fact that all hot bodies glow,* even if we can't see the light with our eyeballs. The Sun, at about fifty-eight hundred degrees Celsius above absolute zero,† glows in visible light. Human beings, at room temperature, glow in the infrared. The universe, about three degrees Celsius above absolute zero, glows in microwave/radio wavelengths, and for a long time we were oblivious to its existence.

In 1964, Arno Penzias and Robert Wilson were working for Bell Labs trying to develop early satellite communication. But when they turned on their machines they got an inexplicable interference signal, and the message they received was *not of this world*. Their radio receiver measured a persistent hiss that wouldn't go away no matter which direction they pointed it. They were listening to the microwave radiation of the early universe.

---

*Yes, yes. We can all make childish jokes. Good for you.
†Absolute zero is the minimum temperature possible (-273 Celsius or -460 Fahrenheit), at which point the motions of atoms stop entirely.

In olden times (about ten years or so ago), you would have been able to detect this radiation without any special equipment. When most televisions got their signal via radio waves, about 1% of the static on a channel with no signal came from primordial radiation. Now that everything's switched to digital, you can't use your television to replicate Penzias and Wilson any longer. Not that it would do you any good. They won the Nobel Prize already.

The background radiation is at almost exactly the same temperature no matter where in the sky you look. Almost, but not exactly. If you look in one little patch of the sky rather than another, the radiation appears a little hotter or colder—albeit at a difference of a few millionths of a degree.

In 2001, the Wilkinson Microwave Anisotropy Probe (WMAP) was launched by NASA with an eye toward measuring the "hot" and the "cool" ripples in the universe, and the map is shown on this page. This is just like an ordinary map projection of Earth that you might see in an atlas, but with the difference that instead of standing on the surface, as you do on a globe, you should imagine that you're standing in the middle, and the map shows what the sky looks like.

What you see before you is a baby picture of the universe. You think kids at portrait galleries are a pain; this picture took five years and about $140 million to take. We guarantee that the universe, unlike your kids, won't grow up before you know it, so what's the point of the picture in the first place?

Take a look at the light and the dark patches. These represent regions where the background is a tiny bit hotter or a tiny bit cooler than average. By "tiny bit" we mean only about one part in a hundred thousand. This detail is of more than just passing interest. Back in the early days of the universe, tiny variations in temperature corresponded to tiny variations in the density of atoms and dark matter. These little excesses are the seeds of galaxies that we talked about earlier.

There's another important reason to look at the background radiation. As we look farther and farther back in time, the universe gets smaller and smaller. This means that everything—the photons, the atoms, and the dark matter—will be pushed closer and closer together, and on average, the universe becomes a more and more energetic place. The contribution of photons becomes especially important as we look back in time because as the universe gets smaller, the wavelengths of the individual photons become smaller as well. We saw this in chapter 6 when we talked about the redshifting of light due to the expansion of the universe. Short-wavelength light means that each photon has more energy earlier in time. Not only does radiation have higher density at earlier times, but each photon has more energy as well.

The end result of these effects is that the farther we look back in time, the hotter the universe gets, and the greater the relative contribution of photons to the total energy density. So, for example, when the universe was one tenth the current size, it was about thirty degrees above absolute zero. When the universe was 1% the size it is now, about seventeen million years after the Big Bang, the entire universe was at room temperature. Before that . . . our story gets interesting.

Why can't we see all the way back to the Big Bang?

Combination (t = 380,000 years)

Way back in chapter 4, we talked about the parts of atoms, and we noted that hydrogen, the simplest atom, is made of a proton surrounded by an electron cloud. It is also by far the most common atom. Today, as in the early universe, hydrogen accounts for about 93% of all atoms.

At room temperature, hydrogen is never found without its electron. But at high temperatures, such as in the interior of the Sun or in the early universe, atoms are continuously bombarded by very-high-energy photons.

Picture Captain Bloodbeard as a proton. No self-respecting pirate would consider himself fully dressed without a parrot on his shoulder, and you can think of the parrot as an electron. The early universe is much like a pitched battle on the high seas. Cannonballs (photons) are constantly whizzing past Bloodbeard, and every now and again— *whomp*, his parrot gets knocked off. Don't worry, they'll both be fine. Of course, pirates and parrots go together like peanut butter and bananas, so it isn't too much longer before another parrot perches on Bloodbeard's shoulder.

Meanwhile, during the battle, the entire scene is one with parrots and cannonballs flying everywhere. In fact, the ships themselves are pretty safe, since the cannonballs typically hit some airborne parrot before doing any damage. But all things, even pirate battles, come to an end. The cannonballs stop flying, and the birds, worn out from all their flying, perch on various shoulders—one bird to a pirate, just as nature intended.

Here's how it played out in the real universe. About 380,000 years after the Big Bang, the universe was a stifling three thousand degrees Celsius and roughly one twelve-hundredth the size it is now. We pick this moment, which we'll call combination,* because it represents the instant when everything changed.

Before combination, the universe was so hot that there were virtually no neutral hydrogen atoms at all, just individual protons and electrons flying around the universe, like our flurry of cannonballs and parrots. All the pieces were there, but they were bouncing around like mad. Photons were constantly colliding, getting absorbed and reemitted. With all these collisions, light couldn't travel very far before being rocketed into another direction. Even if you had been alive 350,000 years after the Big Bang (for instance), you couldn't see very far, because seeing involves light going in a straight line from someplace to your eye.†

After combination, the universe had cooled to a point where the photons could no longer rip electrons off their protons, and the ordinary neutral hydrogen could form in a hurry. Suddenly every nook and cranny was filled with neutral stuff, and photons had nobody to play with. Photons *love* charged particles, but neutral particles—not so much. Instead, photons fly around forever in a very empty universe until (if they are lucky) some few of them might be intercepted 13.7 billion years later in a radio receiver on Earth or on Tentaculus VII.

Since we can't "see" before combination, everything we know about the very early universe has to be inferred from the remnant radiation that soars around, and anything we can glimpse from the stars, galaxies, and clusters around us today. And if we add a little physical reasoning to these observations, we can do a pretty good job of piecing it all together.

---

*If you're the sort with more than one physics book in the house, or you want to check Wikipedia to make sure we're not lying to you, virtually everyone will refer to this moment as "recombination." This is a bad term as far as we're concerned, since the "re" part implies that this is somehow happening again, instead of (more accurately) for the first time.

†And you would have been horribly, horribly burned to death.

Shouldn't the universe be (half) filled with antimatter?

They say that a little bit of knowledge is a dangerous thing, but in this case, you have just the right amount. We're going to remind you of two important facts, and then we'll exploit the hell* out of those two facts to describe events in the early universe. As a reminder:

1. $E = mc^2$

2. If you smash a particle into its antiparticle, they'll both get destroyed and converted into high-energy photons. How much energy? See fact 1.

If an electron and a positron (or any particle and its antiparticle) can smash into each other and make light, then the reverse can happen as well: photons can collide with each other to make a positron and an

---

*When using *A User's Guide to the Universe* as a supplement to explain the facts of life to your own son or daughter, you may want to edit out some of the saltier language.

electron. Or, for that matter, they could make a proton and an antiproton. There's a catch, however.

These particle creations can occur only if the energies of the photons are high enough. It takes a lot of energy to build electrons, but it takes a heck of a lot more to make protons or neutrons and their accompanying antiparticles, since the masses are so much higher.

But wait! If you were paying attention, the cosmic soup was totally awash with high-energy photons—photons energetic enough to make heavy particles. They're everywhere. In the very early universe, big, heavy particles and antiparticles were constantly being made from scratch: quarks (and antiquarks), muons (and antimuons), electrons (and positrons); you name it. But time marches on, and as it does so, the photons become less energetic, which means that we can make less and less massive particles and antiparticles, until they can't make anything at all. That's pretty much where find ourselves today.

To put some numbers on this timeline, when the universe was about one millionth of a second old, it had cooled to a temperature of about ten trillion degrees Celsius. This is staggeringly hot; far, far hotter than the temperatures that exist even at the centers of stars. Even at these tremendous energies, the photons were already too weak to produce protons and antiprotons, or neutrons and antineutrons. However, two photons smashing into each other were still easily energetic enough to produce lots of other things, including electrons and positrons, and they continued to get created until about five seconds after the Big Bang.

Consider what a prodigy the universe really is. As far as matter creation goes, it did its best work within five seconds of its birth. While the rest of us were screaming or wetting ourselves, the universe had already produced all the matter we could ever need.

There's another subtle point that turns out to be pretty important. When photons collide, they produce a particle and an antiparticle, and a particle and antiparticle completely destroy each other and produce photons. As far as we've seen, there's never an interaction that produces or destroys a particle without an antiparticle. The upshot is that we never create a proton without also creating an antiproton, or an electron without a positron. By this argument, there should *always* be exactly the same numbers of particles of matter as antimatter in the universe.

If you don't see a problem, then we'd like you to explain how the world came to be made entirely of matter. It's not just Earth, either. If the Moon weren't made of ordinary matter, then poor Neil Armstrong would have been a goner the moment he touched the surface in the *Eagle* landing module. The Sun, too, is made of ordinary stuff, as are the rest of the stars in our Galaxy. If they weren't, then the cosmic rays bombarding Earth would have a lot of antiprotons, but they don't.

Couldn't there be galaxies made of antimatter? Perhaps. Except that galaxies collide with each other all the time, and we have never seen an extragalactic collision with the sort of sheer, unbridled energy that would erupt if a matter galaxy rammed into an antimatter one. In short, our universe seems to be made of all matter. Why, if matter and antimatter are always created and destroyed in equal numbers, do we end up with so much extra matter?

First, a confession. We don't know why there's this imbalance, but whatever the process, presumably it happened a very short time after the Big Bang, when energies were extremely high. However, even if we can't explain *why* there's an asymmetry, we *can* explain how big the asymmetry is. Very early on, there were about a billion *and one* protons in the universe for every billion antiprotons, and a similar number of photons. Then, when the universe cooled so that protons could no longer be made, those billion antiprotons got destroyed, taking with them a billion protons, leaving just one for every billion photons, which is the ratio we see today.

What changed between then and now? Why should it be that these days neutrons can turn into protons, but we can't make either from scratch without also producing an antiproton or an antineutron? Why wasn't this true in the past?*

---

*We know, Grandpa. Everything seemed better in the old days.

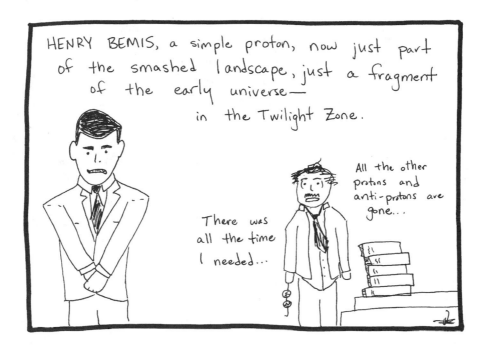

 ## Where do atoms come from?

The Birth of the Elements (t = 1 second to 3 minutes)

We've gotten pretty far from Little Billy's initial question, "Where did I come from?"* but we're now ready to give a somewhat better answer. First we need to tell him what he's actually made of. As you well know, little boys are made of "snips and snails and puppy dogs' tails," which in turn are made of hydrogen, oxygen, carbon, and other atoms.

Taken together, this everyday stuff is known as baryonic matter, which is just a fancy term for everything made of protons and neutrons. When we list the top contributors by mass fraction, we see some old friends:

---

*When confronted, physicists are very good at dodging embarrassing questions.

1. Hydrogen (75%)—one proton
2. Helium (23%)—two protons, two neutrons
3. Oxygen (1%)—eight protons, eight neutrons
4. Carbon (0.5%)—six protons, six neutrons
5. Neon (0.13%)—ten protons, ten neutrons

You don't need to memorize this list, but there is a pretty obvious pattern. With the exception of hydrogen, all of the most common elements have the same number of protons as neutrons. There's even a version of hydrogen, called deuterium, that has one proton and one neutron, and though it's only about 1/100,000th as common as ordinary hydrogen, it will prove pretty important to the story.

If we're any good at our job, we can do more than just make a survey of what's in the universe. We can explain where those numbers come from, and to do that we need to wind our clock all the way to about one second after the Big Bang. Until this point, the leaps we've made have been long compared to, say, the authors' attention span, but as we go back, the jumps will continue to decrease in size (as they must). Think of it this way: about the same amount of important physics happened from the time the universe was one second old to ten seconds, as it did from one billion years to ten billion years.

At one second in, the universe was a piping fifteen billion degrees Celsius, about a thousand times the temperature at the center of the Sun. Even so, at this point the photons were cool enough that they couldn't have made a proton or a neutron from scratch even if they wanted to. But like Bloodbeard and the intrepid naval officer with whom he does battle, there's not as much difference between a proton and a neutron as either would like to think. Turning a proton into a neutron is as simple as smashing an anti-neutrino into it. We also get a positron for free. If you like, we can do the reverse as well. Combine a neutrino with a neutron, and voilà! We give you a proton and an electron to make the charges come out right.

This trick is easier than it sounds, since under normal circumstances if I threw a neutrino at a proton, a neutron, Bloodbeard, or even a light-year of lead, the most likely outcome is precisely nothing. Neutrinos *really* don't like to interact with other particles unless they have to, and whenever they do, it's using the weak force. As any linguist will tell you, the weak force is weak.

However, earlier than one second after the Big Bang, everything was so dense and neutrinos were so energetic that neutrinos and antineutrinos constantly bombarded into protons and neutrons, respectively, turning them into the other and keeping them roughly in balance. "Roughly" is the operative word, since protons are lighter than neutrons, and since nature prefers to keep energies as low as possible, there were a few more protons than neutrons.

Beyond one second after the Big Bang, the distances between particles became too great and the energies of neutrinos became too low to have anything to do with protons or neutrons, and they just went on their merry way, never to be heard from again. But don't be fooled; like the photons from combination, they walk among us still. We just tend to forget them. And that's a shame because they did something very important in the early days by keeping the neutrons and protons in near balance. After the neutrinos went into retirement, the protons, neutrons, and photons engaged in a complicated dance of fusion and fission, in which:

1. Neutrons and protons and deuterium slammed into one another, potentially building heavier and heavier elements.
2. On the other hand, high-energy photons tore the atomic nuclei apart.
3. Any bachelor neutrons broke down* after a while and decayed into protons.

All the while, the universe got more and more diffuse, and cooler, making all of this a race against the clock. When the dance started, we had nearly as many neutrons as protons, so that if the atoms

---

*Just like with real bachelors, some neutrons will end up in their socks and underwear, tired and defeated, drinking warm beer out of cans.

formed very quickly, then all of the neutrons would get paired up and the most common element would have been helium. Helium is the simplest atom with neutrons in it, has equal numbers of protons and neutrons, and is very, very stable. You knew the whole "balance" thing was going to be important, didn't you?

Luckily for us, the protons and the neutrons *didn't* stay in balance with one another, because it would have been a very boring universe. Try making anything out of helium. Good luck with that, pal.

We'll ruin the surprise by noting that we got more than helium out of the Big Bang. The main reason is that this whole process took a few minutes, during which many of the neutrons decided that they'd be better off as protons. They decayed and never looked back. There simply weren't enough neutrons to go around, and remaining protons had to go stag. That's why we have so much hydrogen today.

We've asserted with fair certainty that there is only one baryon for every billion photons out there. We're finally ready to justify our claim. We can measure the number of photons *very* accurately, since we just count up all of the energy coming from the cosmic background radiation. The number of baryons, on the other hand, is harder to come by. We need to start by looking at the blueprints for helium. Elements can't be made all at once, but have to be built part by part. That means that to make helium, we need to start by adding a proton to a neutron and getting a deuterium, hydrogen's huskier brother. Those deuterium atoms (called deuterons when they're not wearing their electrons) can then be fused with protons, neutrons, or other like-minded deuterons. This process goes on for a while until everything cools down and all of the protons and neutrons are locked up in stable elements.

What if we decided to build a universe that was very nearly identical to our own, but it somehow started with exactly twice as many baryons? In the first few minutes, the test-tube universe would be even more crowded than our own was. Deuterium would get made pretty quickly, and what little deuterium we had would smash into protons and other deuterons fairly frequently, taking them out of the equation.

Following this argument to the end, more baryons would have meant *less* deuterium (fractionally) than we have in the "real" universe and somewhat more helium.

If we fine-tune the initial conditions, the chemistry turns out very differently, and by measuring how much deuterium we have, we're basically able to count up all of the baryons in the universe *and* get accurate estimates of other elements as well. So all we have to do is figure out how common deuterium is, and from there, we can compute how much baryonic matter there is. If we look at the very oldest stars and measure the ratio of deuterium to ordinary hydrogen, it is about one in every hundred thousand atoms of hydrogen.

If we pull out the sheet of paper where we've done the full calculation, we get Ω_B in ordinary stuff (not-dark matter) of about 5%. If this number looks familiar, that's because it's the same as the number we get when we count up the mass that we see in stars and gas.

This is amazing! In one fell swoop, we've shown that our model of building elements is, if not correct, at least incredibly accurate, and confirms what we measure from galaxies directly. We understand what's going on one second after the Big Bang and how much ordinary matter there is in the universe. The model even depends slightly (but measurably) on surprising things such as how many different kinds of neutrinos there are. There are three, and these measurements confirm it. From the same model, we're able to accurately predict the amounts of trace elements such as helium-3 and lithium, both of which are observed in exactly the quantities predicted by the model.

But let's not get ahead of ourselves. If only hydrogen, helium, deuterium, and a few other very light elements were created in the Big Bang, then where did everything else come from? Where did carbon and oxygen, the stuff of life, come from? After all, Little Billy certainly could *not* have been made out of the stuff produced in the Big Bang. Everything heavier—carbon, oxygen, gold, and the rest—are created in stars. When the most massive stars explode as supernovas (as we discuss in chapter 6), those heavy elements get strewn out into the universe—ultimately producing the likes of you, me, pirates, and Little Billy.

How did particles gain all that weight?

The Golden Age of Quarks (10^{-12} second to 10^{-6} second)

We've seen a general trend as we look farther and farther back in time. The universe gets hotter and hotter, and the particles get more and more energetic, which usually means that they move faster and faster. For the most part, there were relatively smooth transitions from moment to moment, but every now and again, the changes were quite abrupt.

As a case in point, let's talk about what happened when the universe was about 10^{-12} second old. Before this time, temperatures were so unbelievably high that the Higgs, last seen in chapter 4, couldn't condense to its present-day particle state. As a result, before this moment (if 10^{-12} second could legitimately be called a moment), none of the particles had mass. For some of the particles, such as electrons or neutrinos, acquiring mass didn't end up mattering very much, since they're so svelte anyway. Even after the Higgs particle showed up, they were still trucking around the universe at speeds nearing that of light.

But for others, such as the W and Z particles (the carriers of the weak force), the introduction of mass was a very big deal. Before the universe was about 10^{-12} second old, there was no real difference between the W's and Z's and the photons. This really means that there was no difference between electromagnetism (photons) and the weak force (W and Z particles), and the two were combined into a single "electroweak" force.

What changed? There's a big difference between "has mass" and "doesn't have mass," and the change isn't gradual. In chapter 6 we mentioned that empty space isn't quite as vacuous as you might have supposed. It's chock full of energy, and constantly involves the creation and destruction of particles. This "vacuum energy" causes the Casimir effect and may even cause the acceleration of the universe today. It's also the canvas on which all of the particle interactions take place. At about 10^{-12} second, the *vacuum* changes from a high-energy state to a lower one, which is why the rules of physics seem to change as well. That's how all of the W's, Z's, and Higgses know that they are supposed to have and/or impart mass. When the vacuum drops from one state to another, some of the symmetry that we seem to love so much is lost, and the weak and electromagnetic forces become separate.

This splitting is a general theme in the evolution of the universe. Today there are four distinct forces of nature, but these categories are kind of messy. In chapter 4 we noted that one of the big hopes of physics is that there is a single Theory of Everything that would allow all four forces to be explained by a single law. Einstein spent most of the latter part of his career trying to "unify" all of the fundamental forces understood at the time (only gravity and electromagnetism), but with little success.

While we have a really good theory that combines electromagnetism and the weak force, we're on less solid ground trying to combine the electroweak force with the strong force. We don't know much about how a Grand Unified Theory might work, but we assume that these three forces would be unified at much higher energies than those available at 10^{-12} second after the Big Bang. At even earlier times, it's hoped that the Theory of Everything would unite all four.

But let's not jump the gun. For the moment we can simply ask what the universe was like at about the time when electromagnetism and the weak force broke up.* To answer that, we need to clear up a lie of omission. We've been referring to the idea that there is an asymmetry in the universe, and that at some time, a billion and one protons were created for every billion antiprotons.

That never happened.

There never was a moment when protons and neutrons were produced in great abundance. When the universe was 10^{-12} second old, it was far different than it is now. At that time, the quarks were so energetic that they couldn't even be contained in protons and neutrons. It remained that way until the universe was about a millionth of a second old, by which point the universe had cooled to the point where the quarks would never again be found outside of neutrons or protons.

What all this means is that to some extent we've been worrying about the wrong problem. Instead of focusing on why there was one

---

*Feel free to think of the Higgs particle as the Yoko of the early universe.

extra proton for every billion antiprotons, we should have been focusing on why there was one extra quark for every billion antiquarks. In this way, we just keep pushing the question of Little Billy's origin earlier and earlier.

Is there an exact duplicate of you somewhere else in time and space?

Inflation (t = 10^{-35} second)

The time before the quark epoch was exciting, and also unbelievably confusing. The temperature was so high that quarks, electrons, and neutrinos could all be made easily by high-energy photons. It doesn't make sense for us to worry about which particles were doing what at this point. The particles were all being created and destroyed so quickly, they all were essentially the same. While the universe's primordial ooze was pretty much uniform, and not all that exciting to talk about, we still have some early mysteries worth considering.

Mystery 1: The "Horizon" Problem

The cosmic background radiation is all, for the most part, at the same temperature. Two different spots, each on opposite sides of the sky, might differ by only about a part in a hundred thousand or so. Maybe that doesn't seem like such a big deal, but perhaps a saucy analogy will bring the problem into greater clarity. Consider the pirate captain about to take a soak in his bath. Imagine he fills his tub from two different faucets, one at each end. The faucet on the starboard side fills with cold water, but from the port side, the water comes out piping hot. Do you suppose that if the captain fills his bath and steps in, he'll find himself lukewarm all over? No. His head will be hot and his feet will be cold. The water doesn't equilibrate instantly.

All of the temperature calculations we've been doing are based on the simple proposition that when the universe was half the size it is now it was twice as hot, and so on. But that means that everywhere needed to start at nearly the same temperature. Just like Bloodbeard's bath, if different places started off at different temperatures, they need a little time to mix. The very fastest that they can transfer heat from place to place is at the speed of light, and there simply hasn't been enough time.

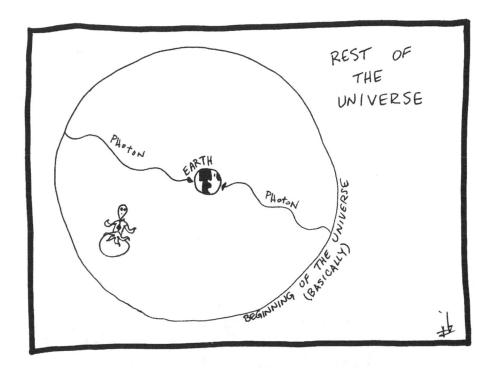

"But wait!" you might object. "We've had almost 14 billion years! Surely that's enough time for everything to get all blended." This is true, but you forget one thing. The light shining on the North Pole and the light shining on the South Pole are coming from patches of space that are very far from one another.

We can even anticipate your next objection: "In the first moments of the Big Bang, all points were *very* close to one another. Surely the two spots could have mixed at that time?"*

---

*When did you get so smart? Not that we're complaining, mind you.

Nice try, but no. While it's true that they were close together, the universe was very, very young at the time. The youngness beats the smallness. We know that your gut reaction is just to assume that the temperature must have been the same everywhere "at the start," but that's only because you're used to thinking about things that had events that preceded them. The Big Bang is the ultimate starting point, so there's simply no reason why the whole thing should have just started out uniform, and you're thinking of the universe as starting from a little pellet. As we saw in the previous chapter, the expansion of the universe simply doesn't work that way. At any time after zero, all points in the universe were some distance away from one another.

Chugging through the numbers, only spots on the sky separated by about one degree of angle, about twice the size of the full Moon, should have had time to mix. Since most of the universe was never in contact with the rest of the universe, why does everything look more or less the same no matter where we look? Why are galaxies in the Northern Hemisphere the same as those in the Southern?

Mystery 2: The Flatness Problem

We saw another big mystery in chapter 6 when we talked about the shape of the universe. We noted two things:

1. There is a critical density of the universe that determines its fate and its shape. When we added up all of the contributions of mass and energy—dark matter, dark energy, baryons, and photons—and then divided by the critical density, we found that the ratio of the actual density to the critical density, Ω_{TOT}, is 100%—or at least as close to 100% as we can possibly measure. This result means that the universe is flat.

2. If Ω_{TOT} were even a little different than 100%, then as the universe evolved, the density would either grow very quickly and ultimately collapse (if it was more than 100%) or shrink (if it was less than 100%). To give you an idea, if Ω_{TOT} was 99.9999% at one second after the Big Bang, then it would be less than a billionth of a percent today.

We now come to the second mystery: why is the universe flat, when as nearly as we can tell, it didn't have to be?

The Solution: Inflation

In the early 1980s, a number of researchers were trying to deal with this issue, as well as with the question of how and when the strong and the electroweak forces were unified. The hope was that at higher and higher energies, the forces behaved more and more alike. While we are nearly able to probe the energies required for electroweak unification using accelerators on Earth, we are not yet at the point where we can experimentally test anything about the unification of strong and electroweak. Even the LHC, which is the most powerful accelerator at our disposal, would have to be able to create energies *trillions* of times what it is now capable of to probe a Grand Unified Theory (GUT).

We can speculate, though. At about 10^{-35} second after the Big Bang,* the energies in the universe were high enough that all three of the nongravity forces might have been unified, and the vacuum was in an *even higher* energy state than during the period of electroweak unification. The temperatures we're talking about are so absurdly high that they look made up: about 10^{27} degrees Celsius. Since we don't have a single model for a Grand Unified Theory, we're a little fuzzy on the details, but if it's anything like the end of the electroweak epoch, something strange happened when it ended.

In 1981 Alan Guth, then at Stanford, proposed that this "something strange" was "cosmic inflation," and at first blush it will sound ridiculous. Shortly after the strong force broke away from the combined electroweak force, the inflationary model says that the universe underwent a period of exponential expansion—increasing in size by a factor about 10^{40} or so in a tiny fraction of a second.

That's the basic picture of inflation, but explaining *why* we think this is a viable model of the early universe is another matter. You might think that exponential growth seems completely unrealistic.

---

*To put this in perspective, during such a short period of time, even light could travel only about one trillionth the way across an atomic nucleus.

Well, don't. Remember that our own universe is starting to undergo exponential growth even as we speak. It's a little thing you know about already, called "dark energy."

You also might be concerned that such a rapid expansion violates the principle of special relativity, since nothing is supposed to travel faster than light, but don't be. We're only concerned about *information* traveling faster than the speed of light. Space is free to expand as it sees fit.

Think of Captain Bloodbeard and his crew trying to spend some of their ill-gotten booty at the mall. Bloodbeard knows that he can't normally outrun his first mate, Mr. Winks, but he finds that if runs up the escalator, he seems to be flying. He's moving way faster than Mr. Winks could ever run. Imagine the captain's surprise when Mr. Winks also runs up the escalator and easily speeds past him.

The same is true in an expanding universe. Particles might seem to be moving faster than light, but that's only because the universe is expanding under them. If you had been a subatomic particle in the early universe, you *still* wouldn't have been able to outrun a light beam. This fact is no less true during inflation than at any other time.

A bigger question might be why this expansion would start in the first place. The idea is that when the strong force decouples, it produces what's called a "phase transition" in the universe. You can think about this effect as a sudden transformation, similar to what happens when you raise ice above zero degrees Celsius and it melts. It's also very similar to the changes when the electromagnetic and the weak forces decouple.

The idea is that during inflation, the universe was filled with something known as the inflaton field.* In many ways, this field is similar to the Higgs field that controls masses today, and relates the electromagnetic and weak fields. Because the inflaton expands just like dark energy, it shares a number of important properties. One of the most important is that like dark energy, the inflaton field doesn't lose energy density as it expands. This is a very important part of the equation, since we've already seen that ordinarily, a big expansion means that the universe should do a lot of cooling, and everything in the universe should instantly freeze.

---

*No, this isn't a misspelling. It's like all the other particles we've seen: electr*on*, phot*on*, inflat*on*.

But the inflaton field is like a giant battery, and once inflation is done, all of the energy is released and the universe gets recharged. Everything becomes nice and toasty, and it's as if the cooling never happened.

You're right to have your concerns. However, we assure you, we wouldn't have introduced inflation unless it was necessary to explain the observational mysteries in the universe. Remember the horizon problem, in which we didn't understand how different parts of the sky got mixed? Inflation solves this problem very simply. Even though there's very little time before inflation, a little patch of the universe will have time to equilibrate to the same temperature throughout. Then the little patch explodes to such a gargantuan size that it includes the volume of the entire observable universe.

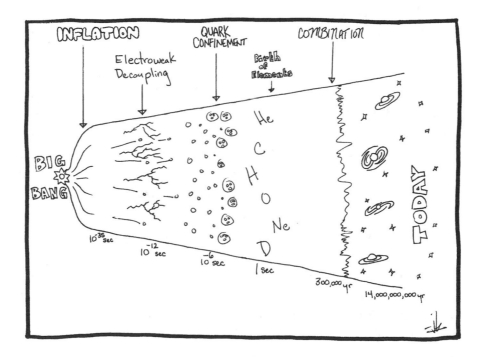

Inflation also explains the flatness problem. This solution is even more intuitive. Imagine blowing up a balloon to huge proportions. Even though the balloon is "really" a sphere, to an ant, or a human, or a galaxy sitting on the balloon, the surface will look flat. In other words, maybe our universe isn't exactly flat, but if it isn't, it's close enough.

Does this mean that there's an infinite amount of matter in the universe? After all, we said that a flat universe was infinite. Since there's a certain amount of matter everywhere, and an infinite amount of space, the natural implication is that there's an infinite amount of stuff in total.

People get concerned about this idea because when you start introducing concepts such as infinity, their minds race to ideas such as, "If there's an infinite amount of space, then there's an infinite amount of matter, and then somewhere in the universe, there's another me," and then they don't feel very special.

In our minds, you are *very* special, even if the universe doesn't think so.

We've been referring to *the* period of inflation as though there were only one, but the truth (according to the model, at least) is that the universe may have branched off many, many times—perhaps infinitely many. Each little patch of space underwent inflationary expansion, and new space got created faster than the inflation in any given patch ended. In other words, the number of universes might have grown without limit. Alan Guth referred to this concept as the "ultimate free lunch."

To keep things clear, we need to distinguish between our universe—everything we see and are in direct contact with (now or in the foreseeable future)—and the "multiverse." The multiverse (one of many terms used to describe the same basic idea) is what we might think of as the Universe (with a capital "U"). The multiverse might consist of many different universes, which either follow one another in time or are separated by space, or simply are never allowed to interact with one another directly.

Don't mistake these different universes with the Many Worlds interpretation of quantum mechanics that we saw in chapters 2 and 5. The different universes of the multiverse are regular old universes that may have properties very similar to our own (or may not), but we simply can't visit them.

Let's suppose there are an infinite number of universes in the multiverse. Quantum mechanics tells us that even if every particular universe is finite, there's only a certain number of ways in which each universe can be arranged (though it's a vastly huge number). This means that somewhere in the multiverse there may be a person identical to you in every way. He or she is reading this very sentence just as you are and feeling just as insignificant. It's humbling, and also just a little bit creepy. It's like you have an infinite number of stalkers. What's more, if the universes really are infinite in number, even our entire universe will be duplicated somewhere.

Is our individual branch from inflation—our universe—infinite? Not necessarily. Inflation doesn't make our universe flat, it just makes it so ridiculously huge that it's as close to flat as we could possibly care about. This also means that technically there isn't an infinite amount of matter, and hence no identical you, at least in our universe. See? We told you that you were special.

Of course, since we *really* aren't sure how gravity and matter worked in the first tiny fraction of a second, all of this is just an educated guess.

 ## Why is there matter?

Perhaps most importantly, inflation might explain why our universe has an extra baryon per billion—and why there's any matter in the universe at all. But first we need to fill in a couple of blanks with regard to matter and anti-matter.

We mentioned before that particles and antiparticles are just the evil twins of one another. Would you even notice if someone came along in the dead of night and replaced all of the quarks in the universe with antiquarks, all of the electrons with antielectrons, all of the neutrinos with antineutrinos, and so on? Physicists call this charge or C symmetry. According to what we've told you so far, everything would be completely the same.

So far we haven't talked about any way in which C symmetry is violated in our universe, but it must be, since it's clearly true that everything is made of matter and not antimatter. There are tiny differences between particles and antiparticles. Neutrinos and antineutrinos aren't exactly the same, as it turns out. Both spin like a little top, but experimentally all neutrinos appear to be spinning clockwise if they are heading toward you, and all antineutrinos appear to be spinning counterclockwise.

This rule seems completely inconsequential, except that it means that swapping all particles to antiparticles does, indeed, make things different. However, we can fix everything if we not only switch particles to antiparticles but also switch left for right. This rule for swapping is known as parity, or P symmetry, and would turn clockwise into counterclockwise and vice versa.

The big question is: if we were to swap C and P, does the physics behave exactly the same?* If it does, then the universe wouldn't distinguish between matter and antimatter, and we'd have no idea why our universe has an excess of either.

Once again, accelerator experiments come to the rescue. At high energy, particles called kaons are produced, along with their antiparticles. For the most part, these kaons and antikaons behave the same, and even produce very similar stuff when they decay. However, in about one case in a thousand, kaons produce different decay products than antikaons. A tiny effect, but it demonstrates that the universe does, indeed, distinguish between matter and antimatter.

The idea is that right around the end of the GUT epoch, energies were high enough to create a hypothetical particle called the X boson. X bosons were very massive and quickly decayed into other particles, including quarks and antiquarks—but *not* in equal numbers. The anti-X, on one hand, should behave in exactly the opposite way and, on average, the two should cancel out. On the other hand, if it turns out that the X bosons behave like the kaons—the antiparticles *don't* exactly mirror the regular particles—then we get a few extra quarks and ultimately a few extra baryons.

So if you want to tell Little Billy where ultimately he (and all of the matter in the universe) comes from, you should tell him that we all come from a violation of CP symmetry in the first 10^{-35} second of the universe.

---

*We're here at FermiLab, where we've secretly replaced all the neutrinos with Folger's Crystals. Let's see if anyone can tell the difference.

What happened at the very beginning of time?

The Beginning, Sort Of (t = 10^{-43} second)

The farther we go back, the hotter and hotter the universe gets, and the more and more speculative everything becomes. We don't know a heck of a lot about the GUT epoch, for example, but since we know how all of the nongravity forces work and have a similar theory for all three, scientists are at least willing to make an educated guess about what a Grand Unified Theory might look like.

On the other hand, we can't even be sure if we're on the right track on how to combine gravity with the other forces or with quantum mechanics. The two theories collide on timescales during which black holes could pop into existence—black holes larger than the horizon scale of the universe. Sound absurd? It should. The timescales we're talking about are about 10^{-43} second—that's forty-two zeroes after the decimal place. This magic number is known as the Planck time, and we can't say much about it with certainty. It's basically a number that pops out of the equations when we toss all of the physical constants in and ask on what scale gravity and quantum mechanics collide.

As we discussed in chapter 4, combining gravity and quantum mechanics is one of the central problems in physics beyond the Standard Model. While approaches such as string theory or Loop Quantum Gravity may ultimately be useful in reconciling things, at this point we just have no way of knowing. If Loop Quantum Gravity is correct, for instance, then not only is there a smallest distance that can ever be measured, there's a smallest time as well. Just as a movie seems continuous until you notice that it's broken down into twenty-four frames per second, it may be that our universe is broken down into frames also.

Even if time and space aren't pixilated on the Planck scale, they still look awfully messy. In 1955, John Wheeler realized that if particles are constantly getting created and destroying each other in the vacuum, then they must have a gravitational field. As a result, on scales smaller than the Planck length, even empty space must be hopelessly deformed and distorted.

He referred to this as the "quantum foam," and if it's real (it's never been seen, of course), then there's a smallest size that anything can possibly be.

Let's ignore all that for the moment and just do the naive thing. We'll assume that we *can* push farther and farther back in time and pretend that ordinary general relativity doesn't break down. Just as we realized that space was something that could fold back on itself and be finite, time can be as well. In other words, according to general relativity, there is no time before the Big Bang. The Big Bang was the creation of the universe, and that includes the creation of time. It's very much like trying to ask, "What's south of the South Pole?"

This is kind of troubling, because even though we have no problem saying that the fabric of space expands, and matter can be created from nothing, in both cases we started with *something*. Even during inflation, when the size of the universe increased by a factor of a gazillion, we still had to start with a small patch and just envisioned that it was very malleable. When we made our particles, we started with energy. As a result, when we talk about the singularity of the Big Bang, it's tempting to think of the whole universe as a very tiny, very dense pellet that exploded. The problem is that that picture is totally at odds with how we know physics to work. However, it's not like we actually have a model of how a real universe could be created from an infinitesimal point.

We can't possibly say anything about what happened before, so stop asking.

We mean it. We just don't know.

Okay, if you insist, we might be able to come up with a few guesses.

What was before the beginning?

It's important enough to repeat: general relativity implies that there simply was no "before the Big Bang." As far as Little Billy is concerned, time simply didn't exist. We have some wiggle room, however. Since we don't know what happened before the Planck time with anything even remotely resembling certainty, we sure as heck don't know what happened before the Big Bang. Regardless, we're left with one of two possibilities:

1. The universe had some sort of beginning, in which case we're left with the very unsettling problem of what *caused* the universe in the first place.
2. The universe has been around forever, in which case there's literally an infinite amount of history, both before and after us.

Neither of these is satisfying, and each poses problems that even religions have a tough time with. Consider "In the beginning . . ." at the start of the Old Testament. We're to understand that God created the world. In that case the universe—our universe—has a definite beginning. *However*, God himself is supposed to be eternal. What was he doing *before* he created our universe?

THE DAYS BEFORE TIME BEGAN WERE REALLY BORING.

It's no more satisfying to assert that the universe somehow created itself. We have to figure out some reasonably plausible model for what caused the universe to get started in the first place. As a particularly clever cheat (or theory, if you prefer), in 1982 Alex Vilenkin of Tufts University showed how what we've learned from quantum mechanics might shed light on the how the multiverse popped into being.

First, Vilenkin noted that if we were to somehow start with a small bubble of a universe, two things could happen. If it were large enough, its vacuum energy would cause it to expand and undergo inflation. If it were small, it would collapse. But we learned something important from Mr. Hyde in chapter 2. Once you introduce quantum mechanics, nothing works like we'd expect it to. Remember when Hyde "randomly" tunneled out of a hole in the ground? In the same way, a small universe can randomly tunnel into a larger one. The amazing thing about Vilenkin's model is that even if you make the "little" universe as small as you like, this tunneling still can occur. It even works if the little universe has no size at all. You know what we call something with no size? Nothing.

Prior to the Big Bang, the state of the universe was something that possessed (no fooling) zero size and for which time was essentially undefined. The universe then tunneled out of nothing into the expanding, branching multiverse we've already seen. The problem is that the "nothing" that the universe popped out of wasn't *really* nothing. It had to somehow know about quantum mechanics, and we've always been taught to think that the physics is a property of the universe. It's troubling to think that the physics existed before the universe did, or, for that matter, before time did.

Of course, this detail is the basic problem with any definite origin for the universe. Somehow all of the complexity had to be created from nothing, and it's difficult to reconcile that.

The other possibility seems equally troubling. The multiverse might literally be eternal—or at least have an infinite history. We're not going to get any further into the philosophical or theological implications here. However, we *can* ask the question of how an infinite multiverse might work.

Infinite Multiverse Scenario #1: The Universe Gave Birth to Itself

If Little Billy was uncomfortable with the implications of "the beginning" before, or even if he thought he had gotten some insight into where all this was going, the phrase "The universe gave birth to itself" is enough to turn anyone off the idea of asking a physicist for clarification in a biological discussion.

The origins of the universe are still fair game, though. In 1998, J. Richard Gott and Li Xin Li, both then at Princeton, proposed a possible variant in which the multiverse arose from what can only be described as a time machine. Gott and Li showed that it was possible to solve Einstein's equations of general relativity in such a way that a multiverse started off going around and around in a continuous loop,[*] and that that loop could serve as the "trunk" of a tree that sprouted, giving rise to our own universe. Since a picture says a thousand words, let's illustrate with their own figure.

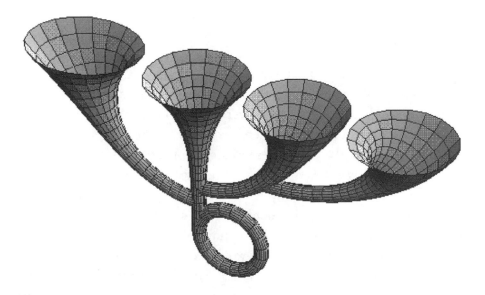

The way to read this image is that for the most part, time travels from bottom to top, and that everything begins with the little loop at the bottom. That is the origin of the multiverse. This means that the multiverse has no beginning, since the loop goes around and around infinitely.

Consequently, we can talk about the "time after the Big Bang" as the time after the loop sprouted off into the future and a universe was born. You'll also notice that there isn't just a single horn coming out of the initial time loop, but many. This image is totally consistent with the multiverse view of inflation that we've already seen.

Infinite Multiverse Scenario #2: This Is Not the First Universe

The possibility exists that the universe might ultimately collapse on itself, a possibility we discussed and almost immediately rejected. The allure, from our current perspective, is that if the universe were somehow to end in a big crunch, then maybe what's really happening is that the multiverse is just a series of expansions and contractions, on and on for infinity, and our universe is just one in an infinite series.

The problem with this (besides the fact that there is too little stuff in our universe to make it collapse again) is one of disorder. As we saw in chapter 3, the universe loves disorder. If you've ever stacked soda cans, there's only one way to stack them, and that's straight up. But if you knock them over, they go everywhere. There are more ways to destroy a soda can tower then there are to build one, and as time goes on, the universe finds ways of destroying all other forms of order, too.

If our universe was the result of a series of expansions and collapses, then our Big Bang occurred billions or trillions of years after some beginning (and what caused that?), so it would have had a very long time to get disordered. But it isn't. Looking back, our universe was very smooth, and in a very high state of order. This theory wouldn't solve the problem at all.

But in recent years, there have been a number of new cyclic models that allow an eternal multiverse to exist. In 2002, Paul Steinhardt, of Princeton University, and Neil Turok, of Cambridge, devised a model that exploits the extra dimensions found in string theory. As we saw in the previous chapter, string theory supposes that our universe might not be three-dimensional at all, but might have as many as ten spatial dimensions. Our own universe might simply live on a three-dimensional brane that is floating through the multiverse, barely interacting with the other universes.

However, the different branes (universes) could interact gravitationally. In this model, the dark energy that accelerates the universe isn't a real thing at all, but just a remnant of the gravitational attraction between branes,* and the dark matter is just ordinary matter on the other, nearby brane. Occasionally the branes collide with one another, which would set off "Big Bangs" within the different branes and then everything would proceed as we've already seen.

These models are extremely elegant, and have the added bonus of not requiring inflation to explain the flatness and horizon problems. They also deal with the whole "increase of disorder" problem in a really novel way. In cycle after cycle, the branes get more and more stretchy, which means that the disorder gets spread out over a larger and larger volume. The local patch that we call our universe, however, is just a small patch of the brane, so we seem to start nearly from scratch at each go-round.

It sounds great, but a big problem is that these models require string theory to be correct, and on that the jury is definitely still out. There are a number of situations in which the universe might undergo a series of contractions and bounces, and string theory is only one possibility. If Loop Quantum Gravity is correct, for instance, then if you tried to run the movie of the universe, you would get stuck at the Planck time—the universe literally couldn't get any smaller or younger than that. As a result, the time would reverse itself automatically. To put it another way, the *natural* solution would be that the universe is eternal.

At the end of the day, the Big Bang theory has the same basic problem as evolutionary theory. Both do a nearly perfect job in explaining how the universe (or life) changed when it first came about, but neither can explain how things *really* got started in the first place. You can't really fault a theory for not explaining absolutely everything, but that doesn't mean that we're not curious. Little Billy's explanation might end with the words "and we don't know where you came from, either."

---

*"Gravitational attraction between branes" is also commonly referred to (at least in the scientific community, where looks don't matter) as "love."

Extraterrestrials

"Is there life on other planets?"

Physicists really do deal with some mind-numbingly difficult questions. Already we've talked about the beginning of time, the end of time, and all of the time in between. We've tackled the vast stretches of space and what makes matter. In our discussion of quantum mechanics, we rubbed up uncomfortably with what is destined to be the biggest question ever: free will versus determinism. A constant stream of weirdness pervades the field, and it's sometimes the safest scientific policy to just lower your head, plug through the calculation, and peek at the answer afterward.*

At the same time, there tends to be some sort of public perception that by thinking about physics on the scale of the universe, we might have some sort of special insight into the true nature of reality, or whether we're alone in the universe. Things that, when asked, make a guy blush and remember that he has a calculation to plug through. The big esoteric questions aren't easy to brush off. Newton, famously, was both the greatest physicist of his (or arguably any other) generation, and was a devout Christian. In between inventing physics and calculus, he still had enough time to ponder

---

*Unfortunately, as a general rule, only the odd-numbered problems are answered in the back of the book.

how many angels could dance on the head of a pin. Applying physics to unphysical questions has a proud heritage, which means that when someone asks us if we believe in aliens, it's not enough to feign ignorance. We'll feign knowledge instead.

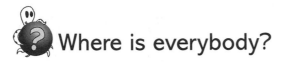 ## Where is everybody?

Let's begin with the obvious. Just because something isn't a physics question doesn't mean that we don't have something interesting to add to the conversation. For instance, "Have we ever made extraterrestrial contact?"

The simplest answer is that since we're not conspiracy theorists, and therefore don't believe in sneaky rendezvous at Area 51, we're pretty sure that UFOs have never crash-landed on Earth. Sure, we *want* to believe, but even so, we would be extremely surprised if we've ever been visited.

Humans have only been broadcasting signals out into space for about sixty years. Aliens wouldn't want to visit us unless they had detected suspicious-looking signals coming from Earth and subsequently wanted to check out where they were coming from (although this desire would probably fade if they actually watched the television they were receiving). Assuming that they set off as soon as they saw the signals, the fastest that they could get here would be just shy of the speed of light.

If any aliens were to visit us, they'd have to be within about thirty light-years or so of Earth. There are about four hundred such stars within the requisite distance, but so far we haven't seen direct indications that any of them contain Earth-like planets, let alone life, and certainly not intelligent life. What's more, since our signals are extremely weak, it's unlikely that an alien civilization would have detected us even if they could be bothered to look for us in the first place.

However, it's a big enough universe that it feels as though there *must* be some other civilizations out there. Enrico Fermi, one of the greatest physicists of the twentieth century, encapsulated the basic problem as follows: Consider the enormous number of stars out there. Unless there

is something staggeringly special about us, odds are that some of those stars—perhaps many of them—might have eventually developed intelligent life. Then, and this point is critical, many of those intelligent civilizations will have spread to other planets. If our experience here on Earth is any indication, people (or peoplelike aliens) can spread to every habitable nook very, very quickly. Since the universe is so very old, it seems like it should be jam-packed full of intelligent creatures, and it's likely that we should have been contacted many times over. As Fermi put it, "Where is everybody?"

Fermi played a bit fast and loose with his numbers, and was perhaps a bit unduly optimistic about the prospects of developing faster-than-light travel or colonizing other galaxies. Still, Fermi's paradox sets the stage for trying to use our knowledge of astronomy and physics to figure out the odds that there are aliens out there with a ticket for Earth. Given what we know about our Galaxy, what are the odds that there are other intelligent beings in it?

The simplest approach would be to take long, repeated observations of many, many nearby stars. In principle, a super-civilization who wanted to advertize their existence to the outside world would send out radio signals with recognizable numerical patterns in them so that other intelligent civilizations could detect them. We, being somewhat less advanced, could only receive the signals; the power necessary to transmit over interstellar distances is well beyond our capabilities. If this scenario seems somehow familiar, it should. It's the basic premise of *Contact*, written by Carl Sagan in 1985 and later made into a perfectly watchable movie starring Jodie Foster.[*]

While the part about us actually making contact is wishful thinking, the science behind the search is not. Since the 1960s, there has been a very active collaboration known as the Search for Extraterrestrial Intelligence (SETI), whose aims are pretty much spelled out in the name.[†] Not to spoil the surprise, but so far the search for ETs hasn't produced anything to phone home about.

Could Aliens Come to Visit (If They Wanted To)?

Imagine that eventually SETI discovers an alien civilization, and—what luck!—they're virtually in our backyard. Suppose we then wanted to mount a manned expedition to visit them on Alpha Centauri, about 4 (actually 4.3, but who's counting?) light-years from Earth. Is this the sort of thing we could do? Practically speaking, no, but it can't hurt to speculate what a little sci-fi–grade engineering might be able to do for us.

We can't use warp drives to travel faster than light, because that's just nuts, and don't even get us started on how impractical it would be to set up a wormhole. We also can't just accelerate instantly up to 99% the speed of light, even if we had the technology—the g-forces would kill us! Let's say

---

[*] One of many instances in which the movie industry has helped out the astrophysics community by suggesting that we are peopled by smoking hotties.
[†] This project hasn't had government support for some time, and thus, since 1999, has largely relied on individual computer users to process the enormous amounts of data produced by the telescope arrays. If you're interested in helping out, check out "SETI @ Home."

our spaceship has only one g of acceleration. We'd be riding in cool comfort. For the first half of the trip, we'd be thrown toward the back of the ship, but because of our rate of acceleration, artificial gravity would feel like Earth-normal. For the second half, during which we slow down, the *front* of the ship would become "down." There's also the issue of energy. Even if our "spaceship" consisted of a single pod with just enough room for a single human,* it still would take as much energy as is currently consumed by the entire United States in a three-month period to get up to speed.

But forgoing those minor technical difficulties, could we make it to Alpha Centauri in our lifetime?

Easily. Crunching through the numbers, it will take only about 1.7 years to travel the first light-year, and only another 1.1 years to travel the next. By the time we're halfway there, we'd be traveling at 94% the speed of light. Of course, at that point we'd have to start decelerating at one g, lest we crash into our destination at nearly the speed of light. Adding it up, the entire trip would take about 5.6 years. It wouldn't make for particularly riveting sci-fi,[†] but it's definitely feasible.

There's a complication: the times here are the times as measured by our friends back on Earth. As we saw in chapter 1, time seems to slow when we travel a sizable fraction of the speed of light. According to someone on the ship, the entire trip would take only 3.6 years—less than the 4-year "minimum" that we'd expect given that Alpha Centauri is 4 light-years away. Make no mistake, we'd still be traveling at less than the speed of light, but our enormous speed distorts both space and time. Because of this time-dilation effect, we could, in principle, visit even more distant stars in our lifetime. The problem is that time passes normally for everybody else, and they might not care to wait for us.

Should we be optimistic? Clearly Fermi thought so, but he figured the aliens could come from any galaxy in the universe. It might be somewhat more realistic to think in terms of just our own Galaxy. We can use a little of that statistical inference we've been touting to try to figure out the

---

*Think US Airways economy class.
[†]Episode 4,182: the intrepid crew finally finishes the game of Monopoly they started in Episode 1,205!

odds of detecting an extraterrestrial civilization. Frank Drake, one of the founders of SETI, formulated a probabilistic approach in the 1960s as to whether intelligent aliens live among us in our Galaxy.

Though it's been rewritten in a number of ways, the Drake equation in its simplest form allows us to multiply together all of the limitations to developing an intelligent civilization:

1. How many stars are there in the Galaxy?
2. What fraction have planets?
3. What fraction of the planets can support life?
4. What fraction of these planets *do* support life at some point?
5. If they develop life, what's the probability that it ultimately evolves into intelligent life?
6. What's the probability that intelligent life will broadcast their existence into space?
7. How long do we expect these civilizations to last?

The first few questions can be answered with a fair amount of precision, but by the time we reach the bottom, your guess is (almost) as good as anyone's. Still, by plying the tools of our trade, we might be able to make some pretty decent estimates.

Let's start with the easiest question, the one at the top. On the face of it, we might expect that there are a lot of intelligent ETs out there. After all, there are about ten billion stars in our Galaxy (the Milky Way), and the typical star lasts tens of billions of years. On average, over the age of our Galaxy, ten new stars form every year, and each star is a new opportunity for a new civilization. What we don't know, however, is how many of these stars might give rise to solar systems like our own. And if there's one thing we're pretty sure of, it's that life needs a planet (or maybe a moon) to call home.

 ## How many habitable planets are there?

When SETI was first established, exactly eight planets were known, all within our solar system and all of them are either too hot, too cold, and/or made of gas. We might be tempted to say that prospects were pretty dim for either finding another civilization or (should we finally trash this planet entirely) finding another place to colonize. It's not that we thought there weren't any planets around other stars. We just hadn't found them yet.*

Things changed in the late 1980s and early 1990s when planets started being discovered left and right. Typically, we find new planets by looking at their star; planets orbit a sun, and the sun technically orbits its planets as well, albeit very slightly. If the planet is massive enough and close enough to its parent star, then the star will appear to wobble a little bit on every orbit, a measurable effect that can be used to infer the mass of the orbiting planets.

We've even started to be able to see some of the newly discovered planets directly. In 2008, groups from UC Berkeley and the Hertzberg Institute in British Columbia took direct images of planetary systems known as Fomalhaut b and HR 8799, respectively. Don't expect to see images of lush beaches and urban skylines. Each of the photos is only one pixel across. Besides, these exoplanets aren't the sort of places you'd like to vacation on, anyway. They're all much more massive than Jupiter and almost certainly made of gas.

In early 2009, NASA launched the Kepler satellite. This instrument will continuously monitor about a hundred thousand stars, and look for signals that planets are eclipsing their host. As the planet passes in front of its star, the light of the star dims by a small amount. Since this effect is periodic, we use the dimming to figure out the length of the planet's year, its physical size, the distance from the star, and other key properties.

So far, we've discovered more than three hundred planets outside our solar system, and doing a rough estimate, it seems that at least 15% of stars have planets, many of them more than one. However, the vast majority of those so far discovered are much more like Jupiter than Earth, and (unless swimming through a giant gaseous sphere of hydrogen appeals to you) aren't the sorts of places where we'd expect life to flourish.

---

*Did you look in your jacket? Well, how about in your pants? No, no. Your *other* pants!

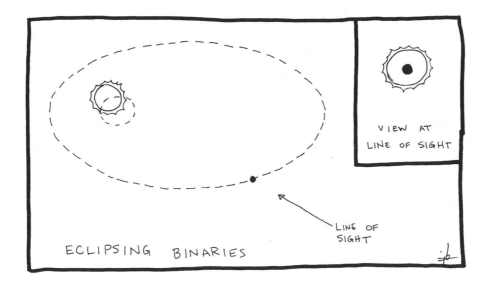

ECLIPSING   BINARIES

VIEW AT
LINE OF SIGHT

LINE OF
SIGHT

What we'd really like to do is find a rocky planet*—"terrestrial," as they say in the biz. This is very tough to do. Since terrestrial planets are much less massive than gas giants, they produce much less wobble in their parent star and so they are much, much harder to detect than their bulkier Jovian brothers and sisters. Still, we're working on it. The hope is that the Kepler satellite will find lots of Earth-type planets—we just don't know how many. It's designed so a lucky alien civilization who'd built their own version of Kepler would be able to detect Earth.

But why wait for Kepler to start sending back results? In chapter 6 we talked about a phenomenon known as gravitational lensing, in which the light of a distant galaxy is magnified and distorted by the gravitational field of a galaxy or a cluster of galaxies sitting between the two of us. Any mass can magnify background light, and so for several decades, astronomers have been monitoring "microlensing" events in which a star or some other object happens to pass between Earth and a distant star. The distant star gets brighter over the course of a few days or weeks, and then dimmer again. Because of these changes, we can detect any sort of mass, including planets, but we have to be very lucky to do so. In 2005, the Optical Gravitational Lens Experiment (so named

---

*A planet populated by Sylvester Stallone clones? We might need to rethink this goal.

on the basis of the crude acronym) saw a tiny extra signal when it was looking at a star. It detected the most Earth-like planet ever seen outside our solar system, with a mass only about 5½ times that of Earth. But we couldn't live on OGLE-2005-BLG-390Lb (as it's been dubbed), since it's about −370 degrees Fahrenheit on the surface.

We're making the assumption that life has to evolve on rocky planets with liquid water because that's the sort of environment that gave birth to us. Maybe that's fair, since we really don't know what sort of other life might be out there. It's entirely possible that the life grows up not on the planet, but on one of the moons surrounding it. There has been a lot of conjecture that Jupiter's moon Europa might have liquid water under the surface. Perhaps life could arise there, or some place very like it in our Galaxy? The only thing we can say is that life doesn't seem to have evolved on the Moon or on any of the other planets in the solar system. Besides, even if it's possible that life could have evolved on a wider range of planets than we're assuming, it doesn't change the basic picture. We'd still expect life to be relatively rare.

Even within our solar system, just being a rocky planet doesn't guarantee that the planet is "class M," as Captain Kirk and the gang used to say. Mercury and Venus are by far too hot, while Mars has no significant atmosphere, and is too cold too boot. Only Earth falls into the Goldilocks zone: just right. Note, too, that all of the planets in our own solar system have nearly circular orbits around the Sun, which means that they don't vary widely in temperature over a year. However, most of the more than three hundred planets discovered outside the solar system have very elliptical orbits, meaning you'd be roasting part of the year and freezing the other part. Neither of these options is very conducive to life.

We have prospects, though. In 2007, a planet known as Gliese 581d was discovered. Although it's about eight times the mass of Earth, it's *almost* close enough to its central star to allow water to melt. Though we don't know whether there *is* any water to melt, or whether there are any greenhouse gases to warm the planet, Gliese 581d remains the current record holder for the best prospect of a life-supportable planet yet found.

So while we've gotten relatively good at finding extrasolar planets, we haven't yet found one that could support life. Going on to the fourth question of the Drake equation, we have to shrug when we read, "What fraction of these planets *do* support life at some point?" Since we only know of one planet that *could* support life, and that planet *does* support life,* it's hard for us to say anything definitive.

There's plenty of reason for optimism. Consider that Earth is about 4.6 billion years old and that life seems to have gotten started after a mere 800 million years. In other words, in the one case we can look at, life seems to have gotten started almost as soon as it could have.

 # How long do intelligent civilizations last?

On Earth, life seems to thrive wherever it's even remotely possible for life to occur. That said, we're not terribly interested in talking to intergalactic bacteria.† We want to get in touch with green-skinned alien babes from Betelgeuse. What's the probability that life develops some sort of intelligence at some point? We don't know, since it seems to only have happened on Earth once, and even then, only in the past couple of million years.

In the popular perception of evolution, there's the idea that all of the monkeys and lungfish and so on have been inexorably evolving toward the highest form: us. Alas, evolution says no such thing. In a real sense, intelligence doesn't necessarily mean that we're more fit for survival, so it's not at all obvious that life will eventually form any sort of advanced intelligence. Our giant calorie-consuming brains and long periods of gestation and childhood helplessness make us very bad investments in the evolutionary lottery. But every now and again (and we have *no* idea how often), those tickets pay off.

So let's take it as read that every now and again a monkey (or gelatinous blob) suddenly decides to invent language and fire and jazz saxophone. How long will those good times roll? Enrico Fermi, as we saw, thought that once a civilization got started it would be

---

*Hint: it rhymes with "girth."
†With all those flagella waving around, it's very hard to interpret the sign language.

both incredibly long-lasting and invasive, meaning that he expected it would have moved in next door by now.

J. Richard Gott of Princeton proposed what he called the Copernican Principle in 1993 to deal with this very question. He made a very simple assumption: you aren't special.* This idea is pretty reasonable, since every time in human history that we've assumed that we *are* special we've turned out to be wrong. Earth isn't at a special place in the solar system; it's just the third planet out of eight, each orbiting the Sun, which is specially placed in the center. The Sun isn't at the center of the Galaxy; it's about twenty-five thousand light-years away from the center. Our Galaxy isn't at the center of the universe;

---

*Despite whatever your mother told you.

nothing is. As we're starting to see from our discovery of extrasolar planets, even Earth—a rocky planet somewhere in the Goldilocks zone—may not turn out to be all that uncommon.

So what if you're an unexceptional specimen within an unexceptional species? It amounts to this: in any given distribution, we'll be somewhere in the middle, though not necessarily at dead center. Imagine a one-hundred-page book containing the names of all of the people who have ever lived and will ever live, all listed in chronological order of birth, but using a very tiny font. You would be *very* surprised to find your name on either the first or the last page of the book. Only 2% of all of humanity would be either at the beginning of civilization or at the end. Are you feeling lucky, punk?

Physicists typically publish a result if they are 95% confident, which means that in our case, you are "average" if you live somewhere between the first 2½% of humanity and the last 2½%. In the beginning of that range, thirty-nine times as many people will live after you than before you, and at the end of the range, one thirty-ninth as many people will live after you as before you.

Let's suppose that the population of Earth has remained and remains constant over the past and future history. We do this just to simplify the calculation, and because it doesn't affect the numbers very much. If we say that "humanity" has already been around for about 200,000 years (kind of an arbitrary cutoff, we know), then with 95% confidence we can say that we'll last between another 5,128 years and another 7.8 million years. It's good to know that doomsday isn't right around the corner, but it's depressing that humanity seems to have an expiration date.

That's not the end of it. Consider a civilization that develops interstellar travel and colonizes the far stretches of the Galaxy. You would be *far* more likely to have been born on a colony than on the original planet prior to space travel, which was the paradoxical part of the Fermi paradox. So either we're the extremely lucky parent generation to an interstellar empire, or both we and our progeny are confined to Earth.

Of course this, like so much else, is merely a probabilistic statement.

The problem with playing these statistical games with the Drake equation is that we don't know most of the factors that go into them to within a factor of ten, and in some cases, to a factor of one hundred. Drake, for instance, plugged in what he thought were reasonable numbers, and estimated that there might be ten other intelligent civilizations in our Galaxy. This hope is one of the main motivations for the SETI project.

But the actual expected number of civilizations could be a hundred, or even a thousand times smaller. That fact, alone, should give you pause. After all, one thousandth of Drake's estimate means that in a galaxy the size of ours, we might only expect to have, on average, about 0.01 intelligent civilizations happily beaming their thoughts into the rest of the universe. But that can't be right! After all, we *know* that there's at least one intelligent civilization, and it's us. The Drake equation can be used as a guess, but beyond that, who knows?

People* are fond of saying, "lightning never strikes the same place twice." In this case, they mean that it's so improbable that intelligent life formed on one planet (Earth) that it's virtually impossible that intelligent life would form both on Earth and elsewhere. But really, it's more accurate to say that lightning never strikes *once.* That is, if we picked a *particular* star and asked whether it was going to form intelligent life, the odds would be vanishingly small. On the other hand, Earth isn't random. If there weren't intelligent creatures on it, we wouldn't be having this discussion.

What are the odds against our own existence?

So here we are—discussing the possibility of our existence. But no such conversation is taking place on the Moon, since there aren't any intelligent Moonites to have the discussion. The very fact that you (or some other intelligent creature) are part of the conversation necessarily means that it needs to be taking place on a world where intelligent life could evolve.

---

*Or, at least, people who think clichés are the greatest thing since sliced bread.

This is even truer in our own universe. We've done a pretty good job so far in discovering a set of physical laws that describes the universe as a whole. The problem, and one that tends to be swept out of sight in most discussions, is that within the Standard Model are literally dozens of numbers that we measure, but couldn't compute from first principles if our lives depended on it. We like to think that there is some underlying set of principles that sets these numbers, but at the moment we just don't know what they are.

We don't know why the electrons, the quarks, or the neutrinos have the masses they do. We don't know why the fundamental forces have the strengths that they do. Small changes in any of these values would change the universe quite dramatically. For example, if the weak force were even weaker, then protons and neutrons would all get converted into helium almost immediately after the Big Bang. Helium, as you may know, is one of the "noble gases," for the simple reason that it refuses to mix with others. In other words, a weaker weak force means no hydrogen. No hydrogen means no chemistry. And no chemistry means no us.

Or to give another example, if electrons were a little lighter than they are now, they would be so easy to accelerate that they would move close to the speed of light, and it would be impossible to form stars. Stars make heavy elements, including carbon, which are necessary to life, so a too-light electron means no stars and no life.

So what if all of these dials and numbers aren't hardwired into the fundamental physics of the universe? What if they really are random? If any of the dozens of parameters were even slightly different, we wouldn't exist! Moreover, given the presumptive need for things such as water (or, at least, complex chemistry) for other intelligent creatures, there wouldn't be any sapience in the universe at all.

The fact that we *do* exist and are here to comment on the utter improbability of our existence is called the anthropic principle, a term coined by Brandon Carter in 1974, who pointed out, "What we can expect to observe must be restricted by the conditions necessary for our presence as observers." This statement is obviously correct, potentially useful, but also largely dismissed among "serious" physicists, many of whom refuse even to discuss it.

The basic idea is that no matter how improbable, if the universe wasn't fine-tuned to support intelligent life, then intelligent life wouldn't be here talking about it. Was it designed for us? Most physicists (including us) don't think so. Is our universe just one of many? Possibly. We talked about parallel universes, but it also may be true that ours is one universe in a much larger multiverse. Perhaps only a small fraction of those universes have conditions necessary to support life, but we, naturally, live in one of them.

Surely there must be more meaning to fundamental physics than the fact that this happens to be a universe that supports life. Probabilistically, though, it looks like we're on our own for the time being.

She Blinded Me . . . with Science Fiction

People often ask us how their favorite television show stacks up from the viewpoint of scientific accuracy. Our answer? Not well. It's not that the writers like messing up facts; made-up science is just more fun. While this list isn't exhaustive, some of the biggest problem areas are listed here.

Nothing travels faster than the speed of light. Space is big, and nobody wants to watch a show that takes centuries. Whether through warp drives, faster-than-light (FTL) travel, or wormholes, nearly every prime-time serial pushes the bounds of science reality.

Because it's expensive (and confusing) to make people fly around aimlessly within their ships and space stations, sci-fi shows generally introduce some sort of artificial gravity. There are really only three ways to do this: spinning the ship (à la *2001: A Space Odyssey*), filling it with magnets, or constantly accelerating the ship, like in our trip to Alpha Centauri. Most shows just throw away the idea altogether, inventing an "artificial gravity" system as a one-fingered salute to science.

What science-fiction community is complete without alien babes? As we've tried to argue in this chapter, alien species are likely to be few and far between. The same is true for "type M planets." Drop a human on a randomly selected planet somewhere in the Galaxy, and he'll asphyxiate, melt, or freeze

within a few minutes. Of course, crushing is also an option. Space is empty, folks.

We'll give a pass to most shows on the issue of building a proper time machine consistent with the laws of physics (see, for example, the design specs in chapter 5). However, almost every show gets the two cardinal rules entirely wrong. First, they're somehow allowed to go back in time *before* they built their time machine, and second, the writers clearly let the characters change their own past.

While we can't judge every show out there (as geeky as we are, we haven't seen *everything*), here's how a few of the popular ones stack up.

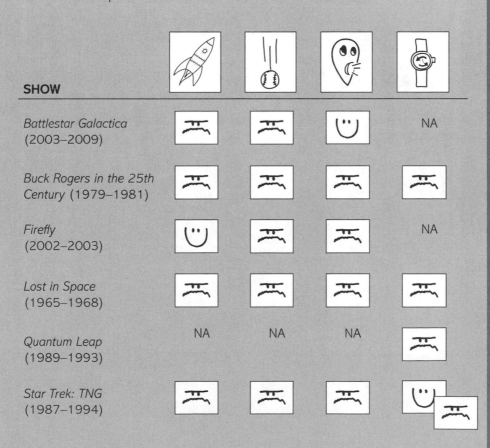

SHOW	🚀	⚾	👽	⌚
Battlestar Galactica (2003–2009)	☹	☹	☺	NA
Buck Rogers in the 25th Century (1979–1981)	☹	☹	☹	☹
Firefly (2002–2003)	☺	☹	☹	NA
Lost in Space (1965–1968)	☹	☹	☹	☹
Quantum Leap (1989–1993)	NA	NA	NA	☹
Star Trek: TNG (1987–1994)	☹	☹	☹	☺ ☹

The Future

"What don't we know?"

If the science fiction of the past is any indication, the planet should already be crawling with cyborgs that turn into fire engines, ablaze with laser swords, and nourished by green, plankton-derivative food substitutes. We have GPS and Tang, but where are our Moon colonies? We can hardly blame sci-fi writers. It's incredibly hard to predict the future. Who, for example, could have predicted that we'd be talking about the possibility of ten-dimensional space and an accelerating universe made up primarily of dark energy and dark matter?

Six Impossible Things (and Six Highly Improbable Ones), before or after Breakfast

They say that with the right attitude, nothing is impossible. "They" are a bunch of idiots. No offense to the motivational poster industry, but there is a fine line between something *seeming* impossible and *being* impossible, in the same way that it isn't easy to grasp the difference between *really, really big* and *infinite*. For instance, it is really, really hard for us to travel at 99.99999% of the speed of light, but it's technically possible. On the other hand, it is absolutely impossible to travel at 100.00001% of the speed of light, even though the latter is only about 130 miles per hour faster than the former. It's not just difficult, not just a challenge—there is no way, no matter how often you cross your fingers, no matter how hard you pedal—it is impossible. Since we've covered a lot in this book, we wanted to give you a handy look-up table in case you ever get into an annoying argument with a pseudo-science type.

SEEMS IMPOSSIBLE (BUT ISN'T)

1. Building a time machine, but only if you were going to anyway.

2. A universe that expands "faster than light."

3. Being in two places at once.

4. There may be an identical "you" in a parallel universe. Not only possible, but deeply creepy.

5. You need to turn an electron around *twice* to make it look the same as when you started.

6. Teleportation: possible, but since current technology limits us to an atom at a time, very inefficient.

ACTUALLY IS IMPOSSIBLE

1. Using a time machine to kill your grandfather. And even if it weren't impossible, you shouldn't do it, anyway.

2. Overtaking a light beam in a fair race. It is, however, possible to use gravity to cheat.

3. Traveling to other dimensions, mostly because it's a meaningless phrase. We're already *in* all of the dimensions, even the tiny ones.

4. Cooling anything down to zero energy. Quantum mechanics always makes your atoms want to jiggle.

5. Escaping from a black hole.

6. Saying with certainty how anything got anywhere.

We've spent a fair amount of time describing the current state of physics, but every now and again we've quietly had to back away from definite claims and sheepishly speculate. Ignorance is a good starting point, and we've identified the limitations in our theories. With the right tools,* perhaps we can address them. With that in mind, strap on your jet packs, because we're going to spend this last chapter looking at some of the big questions that we hope—nay, predict—we'll be able to answer in the next twenty years.

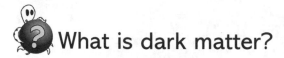 What is dark matter?

Our universe seems stranger than it needs to be. For instance, we found that mysterious dark energy dominates the universe and that most of the remaining mass is completely foreign to us. It's made up of some sort of "dark matter" that *doesn't* interact with light ("dark") but that *is* a source of gravity ("matter"). In other words, the name does nothing more than describe our ignorance. This explanation is only moderately more satisfying than saying that gravity is caused by fairies.

There are some members of the scientific community who are less than convinced that dark matter is a real substance, since we've never detected any dark matter particles. Astrophysicists, for what it's worth, are only doing their job, and suggesting the simplest explanation of their observations—but that doesn't make them right. It wouldn't be the first time that the seemingly "obvious" interpretation turned out to be wrong. The planets and the stars *appear* to move around Earth—an opinion that was generally held until the 1500s, when Copernicus suggested that it's Earth that moves around the Sun.

Some skeptics are so eager to get rid of the idea of dark matter that they've suggested the (nearly) unthinkable—that Einstein and Newton were wrong. A number of theories have been introduced as attempts to stretch Einstein's equations of gravity to make them consistent with observations, without including all of this ugly dark matter. In recent years there has been a lot of interest in theories of MOdified Newtonian Dynamics (MOND).† The basic premise is that on small scales—such

---

*Two liters of Diet Coke, half a dozen neurotic graduate students, and a wheelbarrow of grant money.
†Physicists have made an art out of stretching acronyms to the absolute limits of plausible phonetics.

as those in our solar system and here on Earth—gravity behaves exactly like Newton and Einstein predicted. However, on much larger scales—of galaxies and larger—gravity behaves somewhat differently.

We're not going to defend general relativity just because it was Einstein's brainchild. He was wrong about all sorts of things.* On the other hand, general relativity is extremely "elegant," which in physicists' parlance means that because the equations seem so simple, it's hard to imagine it's wrong. The problem with adopting MOND, as we see it, is that it trades one unexplained number (the amount of dark matter) for another (the scale on which gravity goes from being normal to being "modified").

What's more, you'd have a hard time explaining *all* of the observations that dark matter gets right. MOND does a fine job fixing a problem that's been around for a century—there doesn't seem to be enough mass to hold galaxies and clusters of galaxies together. Since MOND explains away *that* discrepancy, there'd be no need for dark matter, or so the argument goes.

But there's so much more! There also are gravitational lensing observations of the bullet cluster and others that show unambiguously that there are large clumps of matter with no stars or gas associated with them. There are observations of distant supernova explosions that probe the change in the expansion of the universe over time, hinting that there is far more matter than can possibly be explained with baryonic matter alone. Finally, there's all the evidence that the universe is cosmologically flat—which, in turn, makes sense only if 85% of the mass in the universe is dark.

For our money, we're pretty sure that there is a particle out there with the name "dark matter" written all over it—a particle that will be, as the French might say, *le fin du MOND.*

What *Isn't* Dark Matter?

Assume for now that dark matter is real, but wily and evasive. While we don't yet know what dark matter is, we do know a thing or two about

---

*See, for example, the EPR paradox. And despite his warnings, men still wear culottes.

what it *isn't*. It doesn't have any charge, or else it would interact with light. This lack of charge also means that you can't feel it. Everything you've ever touched in your life "feels" like something because the electric fields in your hand are repelling the electric fields in whatever you're trying to touch. No electric fields, and the thing will pass right through you.

In the Standard Model of physics, there are only two known particles that even come close to fitting the bill: the neutrino and the neutron. Unfortunately, the neutrino is far too light, and lone neutrons decay in about ten minutes. Since the universe is somewhat older than that, these are not the particles you're looking for. It would seem that we don't have a good candidate now, but you must remember that physicists are wily, and though there may be a shortage of known particles available as dark-matter candidates, there is no reason why we can't make some up.* Proposed particles include axions, mini black holes, magnetic monopoles, quark nuggets, and many more. Some possibilities, including black holes or magnetic monopoles, have been ruled out observationally or experimentally, though none has come close to being confirmed.

However, many particle physicists believe that WIMPs exist in the universe, and in great numbers. These aren't the runny-nosed lunch money factories you might be thinking of; WIMP stands for Weakly Interacting Massive Particle. Just like the term "dark matter" itself, weakly interacting massive particles just describes most of what we already don't know. Dark matter is, of course, massive, and since it doesn't interact using the strong force or electromagnetism, then presumably it *does* interact using the weak force.†

So WIMP is a good name in that it's descriptive, but a bad one in that it tells us almost nothing at all. The task then goes to theoretical-particle physics to predict what WIMPs are. In this case, "prediction"

---

*Inventing new particles is not as simple as drawing a circle on a coffee-stained diner place mat. Theoretical physicists spend years exploring symmetries, proposing tests in multibillion-dollar accelerators, and, ultimately, drawing circles on soggy cocktail napkins.
†We're being overly glib here. There are a number of dark matter candidates, including axions, magnetic monopoles, and black holes, that *aren't* WIMPs. We'd put the smart money on some variant of WIMP, though.

means more than saying that they exist. A good theory has to tell us how massive WIMPs are, what particles they interact with and how often, and when and how they were formed.

Supersymmetry

The front-runner in our WIMP race follows from a tradition of making up particles that look almost exactly like other particles. As a classic example, consider the neutron. Before 1920 only two "fundamental" particles were known: the proton, which carries a positive charge, and the electron, which carries a negative charge. At the time, scientists were able to measure the nuclei of atoms, and hydrogen, for example, has a charge of $+1$. Helium has a charge of $+2$. The "obvious" conclusion (based on chemistry) was that hydrogen was made of one proton, and helium had two protons, and if those were true, then helium should be twice the mass of hydrogen. In reality, helium has four times the mass of hydrogen.

Drawing on years of training in the physical sciences, Ernest Rutherford recognized that four was bigger than two. He predicted the existence of an electrically neutral particle that was about the same mass as the proton, which was eventually named the neutron. While it seems clear in retrospect, this was a bold prediction. Like dark matter, the neutron doesn't interact with light and thus can't be directly seen. It wasn't until twelve years later that James Chadwick finally discovered the neutron in a lab, and it had all of the properties that Rutherford had predicted.

So there are a number of successful discoveries that stem from physicists saying, "Hmm . . . if we had another type of particle that looked almost exactly like the one I have here, then we'd be in business. Maybe there's a hidden particle, and even though we can't see it for some reason, here's what it would be like." This approach, like in the case of Rutherford's neutron, occasionally brings about new particles that simplify everything.*

---

*On the other side of the coin, there also is a rich history of scientists drinking mercury and storing radioactive samples in their desk drawers.

Physicists love symmetry, as we were made uncomfortably aware in chapter 4. In the Standard Model, there are six different quarks and six different leptons, and each of these groups can be further broken down into two groups of three. In the case of the leptons, there are the three (neutral) neutrinos and the (charged) electron, muon, and tau particles. Moreover, every particle has an antiparticle, with the properties of the two almost exactly identical, but with opposite charges. There are lots of different ways of grouping all the particles, and typically we end up with equal numbers in each group. But that's where the symmetry breaks down. The Standard Model divides all of the particles into two groups:

1. Fermions, the components of matter. Fermions include quarks, electrons, muons, tau particles, and neutrinos, and have the very nice symmetry that we talked about a moment ago.

2. Bosons, the mediator particles. These are the particles that carry the different forces. Bosons include photons, gluons, the W and Z particles, and, if they exist, the Higgs and the graviton.

Counting up everything (including both particles and antiparticles), there are twenty-eight bosons and a whopping ninety different fermions. Don't be too put off by the sheer number of "fundamental" particles out there. Most of them are more or less identical to one another but for some trivial detail, such as the color associated with the quarks.

Still, the fact that there are different numbers of fermions than bosons strikes a lot of physicists as troubling. Why should the particles of matter (fermions) be completely separate from the forces (bosons)? If they are two sides of the same coin, then there should be exactly as many fermions as bosons. This idea is known as supersymmetry, and it means that there are a lot of particles we've never seen. Since they are totally hypothetical, we give them fun, pasta-sounding names such as the gravitino; the neutralino (another candidate for the dark-matter particle); and (our favorite, at least from a whimsical naming perspective) the supersymmetric partner of the W particle, the wino.

These particles behave *almost* exactly like their ordinary-particle counterparts. If supersymmetry really was a perfect symmetry, then a wino[*] should have the same mass as a W particle, a selectron[†] should have the same mass as an electron, and so on. Of course, if they did, then we probably would have made them already in our particle accelerators. Supersymmetry, if it's correct, must be a broken symmetry, meaning that the supersymmetric partners are likely to be much more massive than the originals.

Like the neutron, these supersymmetric particles could decay. As massive particles decay into lighter particles, perhaps the only ones still around are the lightest ones, since they'd have nothing they *could* decay into. Generically this is known as the lightest supersymmetric particle

---

[*]Just so you know, the wino, in and of itself, is *not* a dark-matter candidate. But then again, Rusty's hammered. What does he know?

[†]The extra "s" is for super!

(LSP), and it is thought by many to be the neutralino. If it exists, this LSP might be the dark matter particle we've been looking for.

We'd be remiss if we didn't point out one important fact. To date, there's not a shred of observational evidence suggesting that supersymmetry is correct. It is physics beyond the Standard Model, meaning that it's not technically needed to describe any facets of particle physics we already know of. Still, we've done pretty well in the past by noting symmetries, and there's always a chance it will help us to expand what we know about the universe even further.

How Do We Find Them?

Is dark matter made of an LSP, or is it something else? As long as dark matter is made of some sort of WIMP, discovering them should be relatively straightforward, which is why we're pretty confident they'll be detected in the next few decades. Let's do a quick survey of what we already know. We know to pretty fair accuracy the mass density of dark matter in the universe, so we either have lots of light WIMPs or relatively fewer very massive ones. We know for sure that WIMPs can't be *too* light—less than the mass of a proton—since we've already got lots of accelerators capable of creating light particles, and we haven't seen them yet.

At the other extreme, WIMPs can be only so heavy and still be consistent with cosmological observations. As we've already explained, in the early universe it was vitally important that WIMPs could turn into the ordinary matter we now see and vice versa, which puts a lower limit on how much dark matter and ordinary matter can interact. This lower limit on interaction also sets an upper limit on the mass of a dark matter particle to about forty thousand times the mass of a proton, though this is a huge upper boundary, since most theories predict WIMP masses less than a thousand times that of a proton.

The name of the game is to figure out the mass of the dark-matter particle and the sort of interactions that the particle can engage in, and based on that, we'll see whether those numbers are consistent with supersymmetry, string theory, or something else. Actually getting dark matter particles to experiment on is rough, though, since they literally slip through our fingers. We do have a few options for measuring them, though.

1. *Make them ourselves.* In chapter 4 we spent a lot of time talking about how massive particles such as the Higgs could be created in particle accelerators. Why can't we create dark-matter particles as well? Sure, just like with the neutral Higgs, we won't actually be able to put the dark-matter particles on a table, but the idea is sound. Slam a couple of particles into each other with enough energy and sooner or later we'll produce WIMPs. The measurement of their mass, though, will be based on what we *don't* see. The missing energy in the collisions will be the mass of the WIMP.

2. *You're soaking in it.* We've mentioned again and again* that we're awash in dark matter, but we simply can't detect it directly except by gravity (which is negligible for individual particles) or the weak force (which is normally negligible, period). But we *can* make tubs of liquid, which normally left to their own devices wouldn't do anything. One of the leading efforts, the XENON100 project, uses about three hundred pounds of liquid xenon. Xenon is chosen because it normally doesn't interact with other materials and doesn't decay radioactively. By putting the detectors deep underground, and by carefully testing to see whether cosmic rays are passing through, the idea is that under normal circumstances there shouldn't be any unexplained signal.

   After setting up these tubs and detectors, they just sit and wait for dark-matter particles to whiz by. Every so often, one will hit a proton, and the proton will give off radiation, which is then detected. So far we haven't seen any, but the next generation of detectors is expected to be *much* more sensitive.

3. *Let the universe do the work for you.* The thing about WIMPs is that there are a lot of them,† and they are constantly flying through space. Weakly interacting though they may be, they still interact. What happens when you throw a WIMP into an anti-WIMP?

---

*Why won't you *believe* us?
†See that? Don't be afraid to stand up and be counted.

Usually, nothing. More likely than not, they'll pass right through one another. Every now and again they might do what particles and antiparticles have done since the beginning of time: they'll destroy one another and produce gamma radiation. If we point our telescopes in the right direction, we might see the light from these collisions taking place.

Presumably we'd want to look wherever there is lots of mass. The problem with this approach is that the most obvious places to look include the center of our Galaxy, but there are lots of other things going on there (stuff falling into the central black hole, for example) that *also* produce high-energy gamma radiation. It is very hard to subtract the real signal from the noise, and so far there's been no reliable detection.

In 2008, NASA, in collaboration with the U.S. Department of Energy as well as France, Germany, Italy, Japan, and Sweden, launched the Fermi Gamma Ray Observatory. This space-based telescope may allow us to probe the center of the Galaxy, but also clusters of stars, potential black holes, and other places where dark matter likes to hang out.

Like it or not, dark matter is running out of places to hide.

 ## How long do protons last?

We at the *User's Guide* like to think of ourselves as amateur psychologists.* We assume that people are drawn into physics because they either hope or fear that they'll learn about cataclysms, black holes, and the end of time. You probably slow down to watch accidents, too, don't you?

We won't question your motivations, because, healthy or not, we have the same ones. We've already spent a fair amount of time talking about the evaporation of black holes that await us long into the future, and the so-called second law of thermodynamics, which suggests that as time goes on, the universe will devolve into a tepid bath, unsuitable for structure,

---

*Then again, both of us call our mothers regularly—perhaps compulsively—so we try not to psychoanalyze *too* much.

and wholly uninhabitable to life as we know it. We've even alluded to the fact that the universe seems to be undergoing a never-ending state of exponential expansion caused by dark energy. It will just continue until every galaxy is an isolated island totally cut off from the rest of the universe. Surely the future can't get any bleaker than that, can it?

When you're with a physicist, things can always get worse. What if we told you we have our suspicions that as time goes on, matter itself will slowly boil away and evaporate?

The End of Matter

We know. This topic is really a downer, so the first thing you should realize is that this isn't going to happen overnight. When we talk about galaxies, black holes, and matter evaporating away, we're *not* talking about millions or billions of years. We're talking about time periods a trillion billion times longer than the age the universe is now. And given all of the other bad things that can happen, this seems like it should be relatively low on your list of concerns.

For all practical purposes, when we ask whether matter will decay, what we really mean to ask is whether *protons* will decay. We've already established that given half a chance, a neutron will decay into a proton and some other stuff, but that's only because it's the heavier of the two. The proton is the lightest of the baryons, so we'd expect them to last for a while.

When asked how long, the Standard Model gives us a simple and unambiguous answer: forever. They don't decay because the total number of baryons is supposed to be conserved. Since the proton is the lightest baryon, there's nothing it could decay into.

But if you've learned one thing from this book, it's that the Standard Model clearly doesn't have all of the answers. If a reaction can take place in one direction, then the opposite reaction must be able to happen as well. At some point, way back during the Big Bang, there *must have* been a time when baryons could be made out of nothing. We encountered this problem in chapter 7 when we realized that if baryons and antibaryons were always made in perfect pairs, then they'd annihilate in pairs as well. You are a walking, talking testament to the fact that somehow an excess of baryons over antibaryons got made at some point. Good for you!

The extra baryon production probably occurred at about the end of inflation about 10^{-32} second after the Big Bang, which means it probably has something to do with the unification between the electroweak force and the strong force. If baryon conservation didn't hold then, to some degree, it doesn't hold now.

So imagine you have your own, personal Grand Unified Theory (GUT). One of the first things we're likely to ask is how long a typical proton will last in your GUT.[*] In just about every one of these theories, protons will eventually decay into a positron and another particle, called a pion. The major differences among the various theories involve how long the protons are going to last. This is actually a good thing. It means that if we can figure out how long protons last, we have a major constraint on which of the GUTs are correct—or at least which of them are ruled out.

Looking for Proton Decay

Some of the early models of GUTs predicted that protons should last about 10^{31} years. This is a very, very long time, far longer than the age of the universe, so you might suppose that the physicists who came up with these models were just picking arbitrarily long-lived protons and figured that no one would be able to live long enough to cancel their Nobel Prize check.

Fortunately, we can do better than just putting a single proton on a table and waiting for it to break down. In the 1980s, researchers realized that the thing to do was build giant, underground swimming pools full of ultrapure water.[†] The main purpose for doing these experiments was to see if, left to their own devices, any of the protons in the swimming pool would decay. If they do, the charged particles created from the decay will go streaming through the tank and send off radiation that can be picked up by the detectors on the side. Since there are

---

[*] Ho-ho-ho! You thought we were going to make a joke about belching! Not a chance.
[†] To put things in perspective, the Super-Kamiokande tank is about ten times the volume of an Olympic swimming pool—and all of this about one kilometer underground. That's to protect it from all sorts of other extraneous signals, such as cosmic rays.

lots of protons, it's reasonable to suppose that if we look long enough, one will eventually shed this mortal coil.

We saw something like this in chapter 3 when we talked about the cosmic randomizer. Suppose that the lifetime of a proton really *is* 10^{31} years. It's as if every year the cosmic randomizer rolls a die with 10^{31} sides on it once for every proton in the pool. If any of those rolls come up a 1, the corresponding proton decays. The Super-Kamiokande in the Mozumi mine near Hida, Japan, and experiments of this type have been going on for more than twenty-five years or so, and there still hasn't been a single detection.

This means that we've dramatically underestimated the number of sides on our dice. Instead, we keep adding sides until *not* seeing a decay becomes a reasonable outcome. We've not only ruled out some of the early theories, but we now know that the proton lifetime is *at least* 10^{34} years.

This is good news for us, because it means that we won't spontaneously combust into high-energy particles anytime soon. On the other hand, it's potentially bad news for some GUTs because it means that they can be ruled out conclusively. These days fewer and fewer models are consistent with the longer and longer minimum lifetimes associated with the proton, but many of them are on the order of 10^{35}. Given how close we are to measuring that limit, is it any wonder that we think it likely that we'll measure it soon? Of course, the other option is that we'll have to go back to the GUT drawing board.

 ## How massive are neutrinos?

When talking about the possible dark-matter candidates, we introduced and immediately dismissed neutrinos as a possibility. "Too light," we said. If asked how massive neutrinos actually are, we'd start shuffling around uncomfortably and looking at our feet. The simple fact is that we don't know, and for a long time we thought they might even be completely massless. This isn't the case; as it turns out, our first clues that neutrinos had mass came almost by accident.

Nature's Neutrino Factories

Neutrinos are rascally little devils. Since they only interact using the weak force, we can't put them on a scale, and since they're neutral, we can't manipulate them with an electromagnetic field. We *can*, however, make them in nuclear reactors, and nature's reactors (aka stars) make them in great abundance.

Let us tell you a story. About 160,000 years ago there was a supernova explosion in a nearby galaxy, the Large Magellanic Cloud. Since light takes time to reach us, we only saw the explosion in 1987—and it was one of the most spectacular astronomical events in human history. Along with radiation, lots and lots of neutrinos got released in the explosion as well—enough that huge numbers of them reached Earth. As luck would have it, we had big detectors set up to look for them, and at precisely the same time that we saw the light from the explosions, we detected a spike in neutrinos. In other words, they also traveled here, if not at the speed of light, then as close to the speed of light as we could measure. This detail was but the latest piece of evidence suggesting that even if neutrinos weren't massless, they were very, very light, even by subatomic particle standards.

It might seem fortuitous that we just happened to have neutrino detectors up and running before Supernova 1987A went off. Of course, luck had very little to do with it, especially when we describe what some of the neutrino detectors look like. They're giant underground pools of ultra-pure water. If that sounds familiar, it should. Many of the experiments built to measure proton decay ended up serving double duty* as neutrino observatories.

We can't predict when a supernova is going to go off, so it seems like a bad strategy to simply wait for one in the hopes of capturing its neutrinos. Fortunately, supernovas are not the only neutrino factories out there. Our own Sun produces similar numbers of neutrinos as photons in the course of going about its thermonuclear business. It's just that the photons are much more obvious to us.

The neutrino detection business has been around for a while. By the 1960s there was a fair amount of interest in trying to capture the neutrinos from the Sun, so Raymond Davis of Brookhaven National Labs

*Or single duty, since proton decay hasn't been detected yet.

and John Bahcall, then at Caltech, led the efforts to build (you guessed it) a giant, underground swimming pool. The Homestake Observatory, built in an abandoned gold mine in South Dakota, was essentially a hundred-thousand-gallon tank filled with cleaning fluid.* A neutrino flies in, hits one of the chlorine atoms, turns the chlorine into argon, and the argon decays, giving off light. What could be simpler?

The only problem is that the detectors weren't giving the results that we'd expected. Bahcall predicted about two to three times as many neutrinos as were actually detected in the Homestake experiment. Subsequent experiments, the kinds that used water rather than cleaning fluid, found much the same thing.

Somehow, someone was *stealing* most of the neutrinos! But who?

The Infamous "Neutrino-burglar"

---

*We like perchloroethylene for its heady bouquet, but in a pinch, you can use tetrachloroethene. It still catches your neutrinos, and your guests will never tell the difference.

Identity Theft in the Neutrino World

Up to this point we've been glossing over something that may have occurred to you if you flipped through the "rogues' gallery" of particles in chapter 4. There are three different kinds of neutrinos: electron, mu, and tau. We haven't really distinguished among the different types, but the ones that are created in nuclear fusion are electron neutrinos because there are electrons involved in the process as well. The early neutrino detectors were able to measure only electron neutrinos, making the other two essentially invisible. Maybe the "missing" neutrinos somehow (magic, perhaps) turned from electron neutrinos into something else.

The beauty of physics* is that we can use apparently disparate ideas to unify and explain things that would otherwise be completely incomprehensible. Consider the following three, potentially unrelated, ideas:

1. What we normally think of as the same kinds of particles—a spin-up electron and a spin-down electron, for example—can really act like different particles under some circumstances. The converse is also true. Two particles that we normally think of as different can behave the same under some circumstances. Protons and neutrons, for example, behave exactly the same when only the strong force is involved. If the differences are large enough, we call them two different particles, and if the differences are little (like with the spin-up and spin-down electrons), we call them two different states of the same particle.

2. Many particles aren't in one particular state or another; they're in a combination of two or more different states. We saw in chapter 3 that we can set up an electron so that it's totally random whether we detect it spinning upward or downward. In other words, it was in a combination of upward and downward at the

---

*Note that we didn't say anything about the beauty of *physicists*. That would be much harder to justify.

same time, and each had a probability of being measured when we observed the electron. Quantum mechanics is filled with cases of particles doing two (seemingly) mutually exclusive things at the same time.

3. Particles behave like waves. Back in chapter 2, when we told you this, we neglected a little detail that might be helpful now. If the "wave" is oscillating between two different states, the greater the energy between the states, the faster the wave will oscillate between the two.

Let's combine all of those ideas and make a totally astounding (but correct) leap of faith: neutrinos of different types turn into one another.

Experimentally, we know that we have three different kinds of neutrinos: one that interacts with an electron, one that interacts with a muon, and one that interacts with a tau particle. Just as we can think of an electron as a combination of two different particles, a spin-up electron and a spin-down electron, let's think of neutrinos in the same way. Suppose there are three *different* kinds of neutrinos, numbered 1, 2, and 3, listed in order of increasing mass.

Neutrino 1 is a combination of mostly electron neutrino, combined with a good chunk of mu neutrino, and a smidgen of tau neutrino. Neutrino 2 is a *different* combination, and neutrino 3 is a yet different combination still. Whether we call them three different particles or three different states of the same particle is irrelevant. What is relevant is that the neutrinos are not going to be observed the same way every time. This idea is known as neutrino oscillation, since the neutrinos oscillate among identities: electron, mu, or tau.

Now here's the beautiful part: neutrino oscillation works only if neutrinos have mass, and different masses at that. This idea falls directly out of quantum mechanics. If they don't have different masses, then the energy between the different states is zero ($E = mc^2$ strikes again!), the neutrinos would never oscillate, and we wouldn't observe the phenomenon.

Measuring the Masses

In principle it's relatively straightforward to figure out whether neutrinos oscillate, and hence have mass, although difficult to do in practice, since everything has to be so clean.

1. Get an atmosphere, and constantly bombard it with cosmic rays. Fortunately, we have one of these lying around. The cosmic rays are going to hit air molecules and produce (among other things) mu and electron antineutrinos.

2. Put a big tank of ultrapure water and some detectors deep underground. Since we're waiting for proton decay anyway, we also happen to have a few of these lying around.

3. Count the mu and electron antineutrinos, and see if you get the right numbers.

If neutrinos actually have mass, then on the way from the atmosphere to the detector, a great many of the mu antineutrinos should turn into electron antineutrinos, and so the neutrino detector would measure a deficit compared to what we'd otherwise expect.

In 1998, this very experiment hit pay dirt, and the Super-Kamiokande experiment was the first to detect unequivocal signs of neutrino oscillations, and hence that neutrinos have mass. Subsequent experiments confirmed and put tighter constraints on the mass of the neutrinos.

There are just a few complications, as you might imagine. The first is that these experiments don't just measure the neutrino mass, they also measure the intrinsic mixing among the three types of neutrinos: the amount of mu neutrino in neutrino type 1, for instance. The Standard Model gives us absolutely *no idea* why the mixing among the neutrinos should be the values they are, and we're pretty lucky that the neutrinos are as mixed up as they are. Otherwise it would be very, very hard to measure the fact that they're mixed at all.

The second complication is the fact that it isn't clear why neutrinos have mass at all. The Standard Model originally didn't predict neutrino masses, and lots of recent textbooks on particle physics pretty much assumed that the neutrinos were massless. But given the fact that they have masses, why are the masses so small? The current *upper* limit on any of the neutrino masses is about a million times smaller than the electron, the next-lightest fundamental particle. We still don't have an answer, nor do we have any reason to pick one mass over another.

The third complication is that we don't actually measure the neutrino masses from these experiments. Because of how the math works out, we only measure the difference between the square of the masses of the different types of neutrinos. If we can figure out the mass of one of them, it becomes a simple matter of math to figure out the masses of the other two.

The goal over the next twenty years will be to figure out what the masses are in absolute terms, and to do that, we need to measure the mass of any one of the neutrinos directly. In Germany, an experiment called KATRIN is currently under construction that hopes to measure the mass of the electron neutrino directly.

The design is relatively simple. You start with a big vat of tritium.* Tritium is a relatively rare form of hydrogen that has one proton and two neutrons. It's somewhat unstable, and so after a short while the tritium will decay into helium-3, but the important point is that it also kicks out an electron (which is easy to measure), and an electron neutrino, whose existence and energy have to be inferred. Since we know the total amount of energy released by the decay, and we can measure the amount of energy put into the electron, we *know* that the remaining energy must all be in the neutrino. Since we get to observe lots and lots of decay events, we can measure the minimum amount of energy ever put into the neutrino. The minimum energy it can get is the energy required to make its mass via $E = mc^2$. With this experiment and its successors, we'll be able to measure the mass of the electron neutrino to within 0.04% of the mass of an electron.

With construction scheduled to be completed in 2011, it's our guess that we'll have a good answer sooner rather than later.

 ## What won't we know anytime soon?

There is a long tradition of declaring that the "end of physics" is right around the corner. This scenario seemed to be the case at the turn of the twentieth century, for example. James Clerk Maxwell had successfully described electricity and magnetism, and Newton's gravity seemed to describe everything else. Then quantum mechanics and relativity were discovered, and we seemed farther than ever from unifying physics into a neat, simple, and complete view of the universe. We're still reeling from the discoveries made in the early twentieth century, and, as we've described throughout, some of the mysteries of the quantum world are still waiting to be unraveled.

The point is that it is far too easy to rest on our laurels. The Standard Model of particle physics describes every single particle and interaction—but requires four different force laws and twenty or so free parameters to do so. The Standard Model of cosmology describes the

---

*Only in the world of physics could you wave aside collecting a giant tube of radioactive gas as a "relatively simple" design.

history of the universe, and even gives a plausible history during the dark ages before combination. But amid all of these successes are hidden caveats. We plug numbers into the theory and have no idea where they come from. We aren't able to convincingly unify gravity with the other forces, even though we're able to describe them individually very, very well. And in lots of cases we don't even know what the parameters *are*.

There are other questions we'd love to know more about, but on which there's neither consensus nor a reasonable hope for convergence. Our favorites are:

Is String Theory Right, Wrong, or Neither?

Look up and down, to your left and right, forward and back. Those directions seem to be all that space has to offer. Of course, additional dimensions suffer from the same pesky problems as the tooth fairy and Dr. J's Ph.D.; just because you can't see them doesn't mean that they don't exist.

Over the course of this book, we've introduced a "string theory" on a couple of occasions. It almost seems to be a panacea for all the problems plaguing physics. String theory imagines that all particles are fundamentally exactly the same—just little bits of string. It purports to be a Theory of Everything (TOE), meaning that (if it's correct) general relativity and the weak, strong, and electromagnetic forces will be unified into a single theory. The hope is that in some models of string theory, a compelling explanation for dark matter and dark energy—the source of the exponentially expanding universe—will just drop out naturally.

But there's a price to pay. String theory, in its modern incarnation, describes the universe as having ten dimensions, plus one for time. To understand what these extra dimensions might be like, consider a trapeze artist walking back and forth on a tightrope. The casual observer would say that his motion is confined to be either forward or backward, with no other options available.* A member of the audience, looking at the rope might not even be able to tell that the rope has

---

*We're ruling out the option of "down" for the time being. We assume he's very good.

any thickness at all, and might (if they were a particularly dimwitted yokel) be tricked into thinking that the rope had infinitesimal thickness—that it really is a one-dimensional structure.

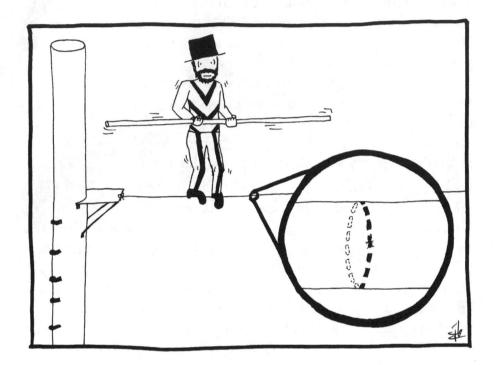

An ant, walking along the rope, harbors no such illusions. He could walk backward and forward along the rope, of course, but he could also walk *around* the rope—the equivalent of one of the hidden dimensions in string theory. Some of the dimensions—perhaps as many as seven— are very, very compact. We may not notice the compact dimensions because we are confined to float along on a three-dimensional brane through a higher-dimensional universe.

Those little dimensions can play a very important role, since quantum mechanics is the major player in this game. What happens if we have a loop of string wrapped up *around* one of the small dimensions? We saw in chapter 2 that if you put a particle in a tiny box (or a tiny dimension), that particle acquires a bunch of extra energy. Normally we'd see the energy expressed by the particle bouncing around. The only problem is that it *can't* bounce around. This means that by reversing our great equation $E = mc^2$, the extra energy becomes the mass of the particle.

The problem is that the energies involved are about 10^{16} times larger than even those produced in the LHC. In other words, there is little prospect of this theory being tested in any experiment that we're going to build for a *very* long time.

Despite popular claims to the contrary, theories in science are never actually proven to be correct. The ones that we accept as the "truth" are simply the ones that have failed to be disproven. The mark of a good scientific theory is that its proponents need to provide a test—or many such tests—that the theory could fail. This concept is called falsifiability, as coined by the philosopher of science Karl Popper, and it is at the core of modern science. This is the major flaw of the so-called theory of intelligent design. You're not allowed to simply state that your theory is correct—even if it explains all currently observed phenomena. As part of your homework, you are required to provide a test—ideally, many tests—that, were your theory to fail them, you'd have to admit that you were wrong, and intelligent design doesn't do that.

How does string theory fare? Consider some recent popular books with titles such as *Not Even Wrong* (Peter Woit) or *The Trouble with Physics* (Lee Smolin), both of which take as their central premise that string theory can be made to be consistent with the Standard Model but that, practically speaking, there is no experiment that could ever *disprove* it. Part of the problem is that there isn't just a single version of the theory. There are an enormous number of possible string theories out there—Smolin estimates 10^{500}, a number so absurdly large that even *Sesame Street*'s own Count von Count would have to reconsider his career choices.

It seems as though, with so many potential alternatives, string theory could be fine-tuned to match just about every possible permutation of physical laws. This result is the opposite of what we'd hoped for. Ideally, we'd want a fundamental physical law that could describe our laws of physics but that wouldn't require us to fine-tune the theory to do so.

As a result of all of this, there's no definite idea of what string theory is, and thus how it can really be tested. As Smolin put it, "There is no realistic possibility for a definitive confirmation or falsification of a unique prediction from it by a currently doable experiment." For our money, we're willing to lay pretty good odds that a definitive test on the dimensionality of our universe isn't coming anytime soon, so even if we don't live in a three-dimensional universe, you should act as though you do.

What Is Dark Energy?

Observationally, there seems to be an unseen but ever-present dark energy in the universe that pushes the universe out toward exponential expansion. The Standard Model even provides a candidate for something that has all of the right properties for dark energy. This is known as vacuum energy, and as we've seen, it has the very real problem that our theory produces numbers that are about 10^{100} times larger than the one observed. We could handle if dark energy were zero—that's a very "natural" number. But this sort of discrepancy just boggles the mind. One of the biggest problems is that theories of string theory and quantum gravity have to be *very* finely tuned to produce the dark energy density we see. To our minds, one of the first tests of a successful TOE is that the dark energy density will drop out naturally.

What's with All of the Free Parameters?

In an effort to describe the general principles governing physics, we've glossed over the fact that there are lots of numbers that just have to be put in by hand. The most natural numbers are simple combinations of physical constants, which means that if we didn't know any better, we'd expect that all of the particles should be the Planck mass or completely massless. They're not, so we might ask why an electron is so much lighter than the Planck mass, and why neutrinos are so much lighter than that. We don't know why the electron has the charge it does, and at the moment, why the strength of the strong force is what it is.

Beyond the scaling numbers, there are tons of parameters that go into the Standard Model, and far more that go into string theories. For example, we mentioned that the various neutrinos can turn into one another, and that there is a mixing factor that tells us about the probability of this transition. Where do these numbers come from? Nobody knows. All told, there are at least twenty free parameters, and that's just in the Standard Model. These are numbers that, as far as our fundamental theories are concerned, could be anything at all.

We hope that in the ultimate TOE, all of the free parameters will be ultimately determined by the theory. But is that true? In the previous

chapter we discussed the conditions in the early universe that were required to give rise to intelligent life. It may be that the parameters really do vary from universe to universe, in which case we never will find a "deeper" reason why the fundamental parameters have the values they do. For our part we find this possibility deeply unsatisfying, and hope it doesn't turn out that way.

Of course, we could be wrong.

This list is, by no means, exhaustive. One of the beautiful things about physics is that there's always a new problem that requires your immediate attention, no matter how many problems you've solved leading up to now. The more questions you answer, the more insight you get into the next question. In traditional *Our Gang* fashion, we get glimpses of the big picture through holes in the fence, and we do our best to piece it all together.

Further Reading

Try as we might, we know that we're not the only game in town when it comes to popular science books. Below are a few that we found especially helpful when writing the *User's Guide*.

Abbot, Edwin A. *Flatland: A Romance of Many Dimensions.* London, 1884. A classic tale of what it would be like to be a two- or one-dimensional creature, with some thought given to how higher-dimensional creatures might live.

Adams, Douglas. *The Hitchhiker's Guide to the Galaxy.* New York: Harmony, 1979.

Bryson, Bill. *A Short History of Nearly Everything.* New York: Broadway, 2003. While Bryson focuses more on classic physics, it's a great overview of some of the great stories behind the science. We "borrow" from his anecdotes liberally in our lectures.

Davies, Paul. *How to Build a Time Machine.* New York: Viking, 2002.

Gamow, George. *The Great Physicists from Galileo to Einstein.* New York: Dover, 1988.

Gott, J. Richard III. *Time Travel in Einstein's Universe.* Boston: Houghton Mifflin, 2001. This is a must-read for anybody wanting to get into the nuts and bolts of practical time-machine construction. We describe Gott's cosmic string time machine in chapter 5.

Greene, Brian. *The Fabric of the Cosmos.* New York: Alfred A. Knopf, 2004. A nice overview of modern cosmology, with particular emphasis on string theory.

Gribbin, John. *In Search of Schrödinger's Cat: Quantum Physics and Reality.* New York: Bantam, 1984.

Kaku, Michio. *Physics of the Impossible: A Scientific Exploration into the World of Phasers, Force Fields, Teleportation, and Time Travel.* New York: Doubleday, 2008.

Krauss, Lawrence. *The Physics of Star Trek*. New York: Basic Books, 1995.

"A Little Bit of Knowledge." *This American Life*. WBEC, Chicago, July 22, 2005.

Mlodinow, Leonard. *The Drunkard's Walk: How Randomness Rules Our Lives*. New York: Pantheon, 2008.

Paulos, J. A. *Innumeracy: Mathematical Illiteracy and Its Consequences*. New York: Hill & Wang, 2001.

Rees, Martin. *Before the Beginning*. New York: Perseus Books, 1998. Not only does this book contain excellent discussions of parallel universes (chapter 2), the multiverse (chapter 7), and the beginning of time, it also has a very nice description of the different interpretations of quantum mechanics.

Rothman, Tony. *Everything's Relative and Other Fables in Science and Technology*. Hoboken, N.J.: John Wiley & Sons, 2003.

Sagan, Carl. *Contact*. New York: Simon & Schuster, 1985.

Smolin, Lee. *The Trouble with Physics: The Rise of String Theory, the Fall of Science, and What Comes Next*. New York: Mariner Books, 2007. Smolin is himself an expert in string theory, and this book serves as both a primer and a scathing critique.

Stevenson, Robert Louis. *The Strange Case of Dr. Jekyll and Mr. Hyde*. London: Longmans, Green, & Co., 1886.

Thorne, Kip S. *Black Holes and Time Warps: Einstein's Outrageous Legacy*. New York: W. W. Norton, 1993.

Tyson, Neil deGrasse. *Death by Black Hole: And Other Cosmic Quandaries*. New York: W. W. Norton, 2007.

Vilenkin, Alex. *Many Worlds in One*. New York: Hill & Wang, 2007. An interesting take on cosmic evolution, including a good description of Vilenkin's own model of the origin of the universe.

Weinberg, Steven. *The First Three Minutes: A Modern View of the Origin of the Universe*. New York: Basic, 1977. A classic picture of our view of the Big Bang. While the story has been extended in the past thirty-plus years, Weinberg's description is still both accurate and intuitive.

Woit, Peter. *Not Even Wrong: The Failure of String Theory and the Search for Unity in Physical Law*. New York: Basic, 2007. The title is self-explanatory, and Woit publishes a blog with the same premise.

Technical Reading

This list consists of stuff that only a specialist (or a masochist) might enjoy.

1. Special Relativity

Einstein, Albert. "On the Electrodynamics of Moving Bodies." *Annalen der Physik* 17 (June 30, 1905): 891–921. This is the classic work in which Einstein derives his theory of special relativity.

Galileo. *Dialogues Concerning Two New Sciences.* Translated by Henry Crew and A. de Salvio. New York: Dover, 1968. Galileo argues in *Dialogues*, his seminal work, that, among other things, Earth went around the Sun, and not the other way around. It also was the origin of the idea of "Galilean relativity," the idea that no experiment can distinguish between standing still or moving at a constant speed.

2. Quantum Weirdness

Barrett, M. D., Chiaverini, J., Schaetz, T., Britton, J., Itano, W. M., Jost, J. D., et al. "Deterministic Quantum Teleportation of Atomic Qubits." *Nature* 429 (2004): 737. One of the first articles on teleportation of individual atoms.

Bohm, David. "A Suggested Interpretation of the Quantum Theory in Terms of 'Hidden' Variables, I and II." *Physical Review* 85 (1952): 166–193.

Bouwmeester, D., Pan, J-W., Mattle, K., Eibl, M., Weinfurter, H., and Zeilinger, A. "Experimental Quantum Teleportation." *Nature* 390 (1995): 575–579. First work on teleportation of photon.

Crisp, M. D., and Jaynes, E. T. "Radiative Effects in Semiclassical Theory." *Physical Review* 179 (1969): 1253. One of several papers showing that Einstein's famous "photoelectric effect" doesn't actually prove that light must be described as photonic particles.

Einstein, Albert. "On a Heuristic Viewpoint Concerning the Production and Transformation of Light."*Annalen der Physik* 17 (1905): 132–148. Einstein's paper showing that light behaves like particles (see Crisp and Jaynes, previous entry, for an interesting addendum). It was for this work, and not relativity, that Einstein won the Nobel Prize in 1921.

Everett, Hugh. "'Relative State' Formulation of Quantum Mechanics."*Review of Modern Physics* 29 (1957): 454–462. Everett describes the Many Worlds interpretation of quantum mechanics. We return to this topic in chapter 5.

Feynman, Richard P. "The Space-Time Formulation of Nonrelativistic Quantum Mechanics."*Review of Modern Physics* 20 (1948): 367–387. Feynman develops his "path integral" formulation of quantum mechanics, in which particles take all possible trajectories.

Goldstein, Sheldon. "Bohmian Mechanics."*Stanford Encyclopedia of Philosophy*, Fall 2008 edition. Edited by Edward N. Zalta. http://plato.stanford.edu/archives/fall2008/entries/qm-bohm/.

Heisenberg, Werner. "Über den anschaulichen Inhalt der quantentheoretischen Kinematik und Mechanik."*Zeitschrift für Physik* 43 (1927): 172–198. This article was the introduction of Heisenberg's Uncertainty Principle.

Huygens, Christiaan, *Treatise on Light,* Translated by Silvanius Thompson. 1678. Reprint, 1945. Chicago: University of Chicago Press.

Riebe, M., Häffner, H., Roos, C. F., Hänsel, W., Ruth, M., Benhelm, J., et al. "Deterministic Quantum Teleportation with Atoms."*Nature* 429 (2004): 734–737. Exactly as the title says: the first experimental teleportation of single atoms.

Schrödinger, Erwin. "Die gegenwärtige Situation in der Quantenmechanik." *Naturwissenschaften* (November 1935). In a brief note, Schrödinger introduces his famous "Cat" thought experiment.

Tonomura, A., Endo, J., Matsuda, T., Kawasaki, T., and Exawa, H. "Demonstration of Single Electron Buildup of an Interference Pattern."*American Journal of Physics* 57 (1995): 117. The double-slit experiment performed with individual electrons.

Vaidman, Lev. "Many-Worlds Interpretation of Quantum Mechanics."*Stanford Encyclopedia of Philosophy*, Fall 2008 Edition. Edited by Edward N. Zalta. http://plato.stanford.edu/archives/fall2008/entries/qm-manyworlds/

Zee, A. *Quantum Theory in a Nutshell.* Princeton, N.J.: Princeton University Press, 2003. A very good technical introduction to quantum field theory for physicists.

3. Randomness

Aspect, Alain, Grangier, Philippe, and Roger, Gerard. "Experimental Realization of Einstein-Podolsky-Rosen-Bohm Gedankenexperiment: A New Violation of Bell's Inequalities." *Physical Review Letters* 49 (1982): 91. Aspect and his collaborators show conclusively that Einstein's interpretation of quantum mechanics was wrong. The universe, at the quantum level, really is random.

Bell, J. S. "On the Problem of Hidden Variables in Quantum Mechanics."*Review of Modern Physics.* 38 (1966): 447. Bell derives his "inequality."

Bennett, C. H., Brassard, G., Crepeau, C., Jozsa, R., Peres, A., and Wootters, W. "Teleporting an Unknown Quantum State via Dual Classical and EPR Channels." *Physical Review Letters* 70 (1993): 1895–1899. The theoretical development of how we might build a practical teleportation device.

Einstein, A., Podolosky, B., and Rosen, N. "Can a Quantum Mechanical Description of Physical Reality Be Considered Complete?"*Physical Review Letters* 47 (1935): 777. This paper introduces the famous EPR paradox.

Greenstein, G. *The Quantum Challenge: Modern Research on the Foundations of Quantum Mechanics.* 2nd ed. New York: Jones & Bartlett, 2005. This very good undergraduate-level textbook describes many of the great issues and experiments in modern quantum mechanics.

Le Treut, H., Somerville, R., Cubasch, U., Ding, Y., Mauritzen, C., Mokssit, A., et al. "Historical Overview of Climate Change." In *Climate Change 2007: The Physical Science Basis. Contribution of Working Group I to the Fourth Assessment Report of the Intergovernmental Panel on Climate Change.* Edited by S. Solomon, D. Qin, M. Manning, Z. Chen, M. Marquis, K. B. Averyt, M. Tignor, and H. L. Miller. Cambridge, U.K.: Cambridge University Press, 2007.

4. The Standard Model

Blaizot, J. P., Iliopoulos, J., Madsen, J., Ross, G. G., Sonderegger, P., and Specht, H. J. "Study of Potentially Dangerous Events During Heavy-Ion Collisions at the LHC."*CERN.* Geneva. CERN-2003-001.

Ellis, John, Giudice, Gian, Mangano, Michelangelo, Tkachev, Igor, and Wiedemann, Urs. "Review of the Safety of LHC Collisions." CERN Technical Document: CERN-PH-TH/2008-136, 2008. The most recent internal review of the possibility that the LHC might create black holes, strangelets, or worse.

Nostradamus, Michel. *Traite des fardemens et des confitures.* (1555, 1556, 1557).

Overbye, Dennis, "Asking a Judge to Save the World, and Maybe a Whole Lot More."*New York Times*, March 29, 2008. One of many examples of public attempts to stop the LHC because of perceived dangers to the world.

http://www.thepetitionsite.com/1/the-LHC

Rutherford, E. "The Scattering of Alpha and Beta Particles by Matter and the Structure of the Atom."*Philosophical Magazine* 6 (1911): 21. Rutherford's discovery of the nucleus of atoms.

5. Time Travel

Einstein, Albert. "Die Grundlage der allgemeinen Relativitätstheorie." *Annalen der Physik* 1916: 49. The original general relativity paper.

Feynman, Richard P., Leighton, Robert B., and Sands, Matthew. *The Feynman Lectures in Physics.* Reading, Mass.: Addison-Wesley, 1971. In 1962, Richard Feynman developed a series of lectures aimed at freshmen at Caltech on all the fundamentals of physics then known. In a sense his lectures were somewhat misguided, as they were far ahead of the students that they were aimed at. However, advanced students, members of the public, and fellow faculty members also attended, and the recorded lectures and subsequent books are among the most interesting reads around for a physicist who knows the math but wants to get a more intuitive feel for the science.

Ghez, A. M., et al. "The First Measurement of Spectral Lines in a Short-Period Star Bound to the Galaxy's Central Black Hole: A Paradox of Youth."*Astrophysical Journal* 586 (2003): L127–L131. One of the first definitive measurements of the black hole at the center of our Galaxy.

Gott, J. Richard III. "Closed Timelike Curves Produced by Pairs of Moving Cosmic Strings: Exact Solutions."*Physical Review Letters* 66 (1991): 1126–1129. This is the technical paper describing the "Gott time machine." For a less equation-ridden description, check out his version in "Time Travel in Einstein's Universe."

Gott, J. Richard III, and Freedman, D. "A Black Hole Life Preserver." http://arxiv.org/abs/astro-ph/0308325 (2003). Gott and Freedman show that the period of time between being mildly uncomfortable and being ripped apart by a black hole is about 0.2 second.

Hawking, S. W. "Black Hole Explosions?"*Nature* 248 (1974): 30.

————. "Chronology Protection Conjecture."*Physical Review D* 46 (1992): 603. Hawking postulates that the laws of physics should not allow the appearance of "closed timelike curves"—that is, time machines. Gott and Li (see chapter 7 on the Big Bang In *A User's Guide*) showed that general relativity does, in fact, permit such a solution.

Matson, John. "Fermilab Provides More Constraints on the Elusive Higgs Boson."*Scientific American*, March 13, 2009.

Morris, M. S., Thorne, K. S., and Yurtsever, U. "Wormholes, Time Machines, and the Weak Energy Condition." *Physical Review Letters* 61 (1988): 1446. Morris and his collaborators develop a model of time machines based on wormholes. Thorne describes this in nontechnical terms in his "Black Holes and Time Warps: Einstein's Outrageous Legacy."

Novikov, I. D. "Time Machine and Self-Consistent Evolution in Problems with Self-Interaction."*Physical Review D* 45 (1992): 1989–1994. While this work isn't the original description of "Novikov's theorem," Novikov uses this paper to go through a number of examples of how time machines enforce a consistent history.

Pound, R. V., and Rebka, G. A. Jr. "Gravitational Red-Shift in Nuclear Resonance."*Physical Review Letters* 3 (1959): 439–441. A test of general relativity from the surface of Earth.

Schödel, R., et al. "A Star in a 15.2-Year Orbit around the Supermassive Black Hole at the Centre of the Milky Way."*Nature* 419 (2002): 694–696. One of the first measurements of a black hole at the center of the Galaxy.

6. The Expanding Universe

Akerib, D. S., et al. "Exclusion Limits on the WIMP-Nucleon Cross-Section from the First Run of the Cryogenic Dark Matter Search in the Soudan Underground Lab."*Physical Review D* 72 (2005): 052009.

Asztalos, S., et al. "Large-Scale Microwave Cavity Search for Dark-Matter Axions."*Nucl. Instr. Meth.* A444 (1999): 569. This is a description of theaxion dark-matter experiment ADMX.

Bondi, Hermann. *Cosmology.* Cambridge, U.K.: Cambridge University Press, 1952. This is generally acknowledged as the first use of the cosmological principle.

Bradac, Marusa, et al. "Strong and Weak Lensing United. III. Measuring the Mass Distribution of the Merging Galaxy Cluster 1ES 0657-558."*Astrophysical Journal* 652 (2006): 937–947. Bradac and her collaborators perform a gravitational lensing analysis on the so-called bullet cluster. In it, they identify giant clumps of matter that don't correspond to mass. This is considered by many (including us) as the first "direct" detection of dark matter.

Casimir, H. G .B. "On the Attraction between Two Perfectly Conducting Plates." *Proc. Kon. Nederland. Akad. Wetensch.* B51 (1948): 793.

Clowe, D., Bradac, M., Gonzalez, A. H., Markevitch, M., Randall, S. W., Jones, C., Zaritsky, D., "A Direct Empirical Proof of the Existence of Dark Matter." *Astrophysical Journal Letters* 648 (2006): 109.

Copernicus, Nicolaus. *On the Revolutions of the Heavenly Spheres.* 1543. Translated by Abbot Newton. New York: Barnes & Noble, 1976. Copernicus conjectured that Earth revolved around the Sun, a position that was later demonstrated by Galileo and ultimately explained by Newton. To this day the Copernican Principle represents (broadly) the idea that Earth (or humanity) doesn't occupy a special place in the universe.

Cornish, Neil, Spergel, David, and Starkman, Glenn. "Circles in the Sky: Finding Toplogy with the Microwave Background Radiation."*Classical Quantum Gravity* 15 (1998): 2657–2670. This work explores the possibility that the universe might just be an infinite spatial loop, much like a torus. By looking for "circles in the sky" (and not finding them), the group showed that if the universe is a torus, it's on a scale much larger than the current horizon.

Hinshaw, Gary, et al. "Five-Year Wilkinson Microwave Anisotropy Probe (WMAP) Observations: Data Processing, Sky Maps, and Basic Results."*Astrophysical Journal Supplement* 180 (2009): 225–245. The WMAP satellite observes the background radiation in the universe and thus presents a picture of the universe as it was very early on. It has phenomenally confirmed our standard cosmological model. This work represents the most up-to-date data release.

Lense, J., and Thirring, H. "Über den Einfluss der Eigenrotation der Zentralkörper auf die Bewegung der Planeten und Monde nach der Einsteinschen Gravitationstheorie. Physikalische." (On the Influence of the Proper Rotation of Central Bodies on the Motions of Planets and Moons according to Einstein's Theory of Gravitation). *Zeitschrift* 19 (1918); 156–163. The "Lense-Thirring effect" is predicted by general relativity and observed by the Gravity Probe B satellite. Basically, it says that a rotating massive body will drag space along with it.

Mach, Ernst. *The Science of Mechanics: A Critical and Historical Account of Its Development.* LaSalle, Ill.: Open Court, 1960.

Perlmutter, Saul, Turner, Michael S., and White, Martin. "Constraining Dark Energy with SNe Ia and Large-Scale Structure."*Physical Review Letters* 83 (1999): 670. One of the first direct measurements of the accelerating universe and thus that the universe is filled with dark energy.

Rainse, D. J. "Mach's Principle in General Relativity."*Monthly Notices of the Royal Astronomical Society* 171 (1975): 507.

Riess, Adam G., et al. "Observational Evidence from Supernovae for an Accelerating Universe and a Cosmological Constant."*Astronomical Journal* 116 (1998): 1009–1038. Riess's group technically scooped Perlmutter's (see previous entry) in the first detection of the accelerating universe.

Rubin, Vera, and Ford, W. Kent Jr. "Rotation of the Andromeda Nebula from a Spectroscopic Survey of Emission Regions."*Astrophysical Journal* 159 (1970): 379. Based on the speed of their rotation, this work represents one of the first pieces of evidence of dark matter in galaxies.

Rutherford, Ernest. "Bakerian Lecture: Nuclear Constitution of Atoms."*Proceedings of the Royal Society A* 97 (1920): 374. One of the earliest discussions to hint at the idea of supersymmetry.

Schmidt, Brian, et al. "The High-Z Supernova Search: Measuring Cosmic Deceleration and Global Curvature of the Universe Using Type Ia Supernovae." *Astrophysical Journal* 507 (1998): 46. Another estimate of dark energy from supernova explosions.

Shapley, Harlow. "Globular Clusters and Structure of the Galactic System."*Publications of the Astronomical Society of the Pacific* 30 (1918): 42. Shapley showed that our Sun isn't at the center of the Galaxy.

Tytler, David, Fan Xiao-ming, and Burles, Scott. "Cosmological Baryon Density Derived from the Deuterium Abundance at Redshift z = 3.57."*Nature* 381 (1998): 207. Tytler and his collaborators measured the abundance of deuterium, which, in turn, allows us to estimate the density of baryonic (ordinary) matter in the universe.

7. The Big Bang

Gott, J. R., and Li, L-X. "Can the Universe Create Itself?"*Physical Review D* 58 (1998): 3501. A model in which the Big Bang can be traced back to a self-perpetuating time loop.

Guth, A. H. "The Inflationary Universe: A Possible Solution to the Horizon and Flatness Problems."*Physical Review D* 23 (1980): 347. Guth's original paper introducing the inflationary picture of the early universe.

Hinshaw, Gary, et al. "Five-Year Wilkinson Microwave Anisotropy Probe (WMAP) Observations: Data Processing, Sky Maps, and Basic Results."*Astrophysical*

Journal Supplement 180 (2009): 225–245. The WMAP satellite took the map of hot and cold spots shown in chapter 7.

Kaluza, Theodor. "Zum Unitätsproblem in der Physik."*Sitzungsber. Preuss. Akad. Wiss. Berlin.* 1921: 966–972. One of several different (independent) derivations of the Kaluza-Klein theory. The idea is that the laws of electromagnetism can be formulated as the properties of a small fourth dimension.

Klein, Oskar. "Quantentheorie und fünfdimensionale Relativitätstheorie."*Zeitschrift für Physik* 37:12 (1926): 895–906. This is the "Klein" in "Kaluza-Klein."

Peacock, John A. *Cosmological Physics.* Cambridge, U.K.: Cambridge University Press, 1999. A very good (albeit technical) overview of cosmology at the graduate level.

Penzias, A. A., and Wilson, R. W. "A Measurement of Excess Antenna Temperature at 4080 Mc/s."*Astrophysical Journal* 142 (1965): 419–421. Penzias and Wilson won the Nobel Prize for their observation that we are surrounded by very-low-temperature radiation—a remnant of the early universe.

Steinhardt, Paul J., and Turok, Neil. "The Cyclic Model Simplified."*NewAR* 49 (2005): 43. Simplified, that is, to the level that ordinary (nonstring theorist) physicists can understand it. The cyclic universe suggests that ours is not the first universe and, in principle, it may not be the last.

Tegmark, Max. "Parallel Universes." *Scientific American* (May 1993): 41–53. Tegmark computes the nearest "duplicate" universe to our own at a distance of about $10^{10^{115}}$ meters away.

Vilenkin, Alexander. "Creation of Universes from Nothing."*Physics Letters B* 117 (1982): 25. Vilenkin describes how quantum mechanics may form a universe out of a random fluctuation.

8. Extraterrestrials

Beaulieu, J.-P., et al. "Discovery of a Cool Planet of 5.5 Earth Masses through Gravitational Microlensing."*Nature* 365 (2006): 623. This was a serendipitous microlensing observation of a star in which a secondary signal was detected. That signal was a planet about 5.5 times the mass of Earth, the lightest extrasolar planet yet detected.

Carter, B. "Anthropic Principle in Cosmology." In *Current Issues in Cosmology.* Edited by Jean-Claude Pecker and Jayant Narlikar. Cambridge, U.K.: Cambridge University Press, 2006.

Gott, J. R. "Implications of the Copernican Principle for our Future Prospects."*Nature* 363 (1993): 315. Gott predicts the duration of humanity, the Berlin Wall, and Broadway plays using a relatively simple probabilistic assumption.

Kalas, Paul, et al. "Optical Images of an Exosolar Planet 25 Light-Years from Earth."*Science*, November 13, 2008. One of the first visual detections of a planet outside the solar system seen by direct light rather than through the wobble of the host star.

Koch, David, and Gould, Alan. *Kepler Mission.* http://kepler.nasa.gov/index.html

Marois, C., et al. "Direct Imaging of Multiple Planets Orbiting the Star HR 8799."*Science Express*, November 13, 2008.

Schneider, Jean. *The Exoplanet Encyclopedia.* http://exoplanet.eu. An up-to-date compendium of all planets found outside the solar system, including references to their discovery.

Tegmark, Max. "Is 'the Theory of Everything' Merely the Ultimate Ensemble Theory?"*Annals of Physics* 270 (1997): 1–51.

9. The Future

Bahcall, John N. "The Solar-Neutrino Problem."*Scientific American* 262 (1990): 54–61.

Bekenstein, J. D. "Revised Gravitation Theory for the Modified Newtonian Dynamics Paradigm."*Physical Review D* 70 (2004): 083509. Bekenstein developed a form of gravity—"modified Newtonian dynamics" (MOND)—that is consistent with relativity, and ideally consistent with all observations without requiring dark matter or dark energy. It is known as tensor-vector-scalar theory (TeVeS), and it's what most people talk about when they talk about MOND. The jury is still out, and for our money, dark matter, dark energy, and traditional general relativity constitute a much more satisfying description.

Bertone, Gianfranco, Hooper, Dan, and Silk, Joseph. "Particle Dark Matter: Evidence, Candidates, and Constraints."*Physics Reports* 405 (2005): 279–390.

Committee on the Physics of the Universe, National Research Council. *Connecting Quarks with the Cosmos: Eleven Science Questions for the New Century.* Washington, D.C.: National Academies Press, 2003. We are not the only ones who have a few unanswered questions about cosmology.

David, R., Jr. "The Search for Solar Neutrinos."*Umschau* 5 (1969): 153. Davis was the head of the Homestake Neutrino Observatory, the one that first detected neutrinos from the Sun, and thus recognized (along with John Bahcall, see previous entry) that there were missing neutrinos.

Distler, Jacques, Grinstein, Benhamin, Porto, Rafael A., and Rothstein, Ira Z. "Falsifying Models of New Physics via WW Scattering."*Physical Review Letters* (2007): 041601. One of many attempts to try to test or falsify string theory using conventional experiments such as the LHC. We're somewhat skeptical because the energies probed by the LHC are far, far lower than those on which string theory becomes important.

Hewett, JoAnne L., Lillie, Ben, and Rizzo, Thomas G. "Black Holes in Many Dimensions at the LHC: Testing Critical String Theory."*Physical Review Letters* 95 (2005): 261603. As with the Distler paper (see previous entry), another attempt to falsify string theory at low temperatures.

KATRIN collaboration. *KATRIN Project Homepage.* http://www-ik.fzk.de/~katrin/index.html

Popper, Karl. *The Logic of Scientific Discovery.* New York: Basic, 1959. This is probably *the* foundational book on the modern interpretation of the scientific method.

Xenon 100 Collaboration. *Xenon100 experiment webpage.* http://xenon.physics.rice.edu/xenon100.html

Index